THE MEANING OF UNITY IN ENERGY CONVERSION SYSTEMS

THE MEANING OF UNITY IN ENERGY CONVERSION SYSTEMS

THE MEANING OF UNITY IN ENERGY CONVERSION SYSTEMS

James F. Murray, III

Published by
A&P Electronic Media
Spokane, Washington

THE MEANING OF UNITY IN ENERGY CONVERSION SYSTEMS

Cover Design: Amazon.com
Editor: Aaron Murakami
Cover Image: Aaron Murakami

Author Copyright © 2019-2020 James F. Murray, III

Publisher Copyright © 2019-2020 A & P Electronic Media

All Rights Reserved, Worldwide. No part of this publication may be reproduced, stored in an electronic retrieval system, or transmitted in any form, or by any means, without the prior, written permission of the copyright holders or the publisher. Unauthorized copying of this digital file is a criminal offense and prohibited by International Law.

Digital and Print Edition Published by:

A & P Electronic Media
PO Box 10029
Spokane, WA 99209
https://emediapress.com

Digital and Print Version 1.00 – Release Date: January 2020

Digital version available at:
https://emediapress.com/jimmurray/unity

Amazon ISBN-13: 9781650183657

THE MEANING OF UNITY IN ENERGY CONVERSION SYSTEMS

Table of Contents

Foreword	1
Introduction	3
The Identification of Motor Losses	7
Types of Efficiencies and Related Calculations	11
Merging Two Efficiencies	14
Force, Voltage and Power in Rotating Components	21
The Dynaflux Alternative	26
Increasing Dynaflux Conversion Efficiency	41
Adhering to Tesla's Advice	47
Conclusion	50
Dynaflux Alternator Patent US4780632A	52
Dynaflux Alternator Patent Application US2013/0187586A1	73

THE MEANING OF UNITY IN ENERGY CONVERSION SYSTEMS

Foreword

I attended my first Energy Science and Technology Conference (ESTC) in 2017 specifically to meet and talk to Jim Murray and Paul Babcock because they built machines they claimed produced more electrical power output than the electrical input power needed to run them: a topic of great interest to me. Unfortunately, Jim was not able to attend the 2017 conference; however, he presented much of the material in this short book at the 2018 ESTC conference. I knew Jim was a very good experimenter from recordings of his previous talks at ESTC conferences, and I was eager to see the content of his presentation in written form so I could go through it carefully.

Jim has been studying, building, testing and thinking about over unity electrical machines for probably more than 40 years. His knowledge of this topic is pretty much unparalleled. When he says that to understand over unity machines, you first need to understand the meaning of unity in energy conversions systems, then you better pay attention. He introduces the ideas of system efficiency and conversion efficiency. Pretty much everybody knows that system efficiency is Power output/Power input x 100%, but I had never heard of conversion efficiency. Jim explains the difference with carefully chosen, easy to follow examples. He then applies these ideas to one of the machines he built, the Dynaflux Alternator, that he received a patent for in 1988.

It seems to me that we are closing in on an explanation of electrical machines that can run themselves. These types of machines seem too good to be true to most people, but they exist. Hopefully it will soon be more widely recognized that this is not only possible, but that it is irresponsible to make electricity any other way. This short book is a good place to start if you are interested in taking a serious look at understanding these important machines.

Jack Hanlon
PhD in EE
Los Alamos, NM

Introduction

The development of the electric motor was a long and arduous process, which probably began sometime in the early eighteen hundreds. However, most research in this direction was severely hampered, in those early days, because of a pronounced absence of sources of experimental electrical currents. But, the advent of the practical storage battery in 1800, originally invented by Allessandro Volta, greatly accelerated electromagnetic research, and quickly resulted in a plethora of electro-dynamic inventions.

The first rotating device driven directly by electromagnetism was conceived by an English inventor; one Peter Barlow in 1822. The invention was commonly referred to as Barlow's Wheel, and in many respects, it was an ingenious contrivance, although not a practical one. But, just a decade later, a Prussian researcher, Moritz Jacobi created the first real rotating electric motor in May 1834.

His device actually developed considerable mechanical output power. In fact, his motor set a world record at that time, which was displaced only four years later by Jacobi himself. His second motor was powerful enough to drive a twenty-six-foot paddle-wheel boat, containing 14 people across the River Neva near Saint Petersburg, Russia. This event took place on September 13, 1838 and was greatly celebrated at the time.

This technological achievement was not eclipsed until 1840, when it was finally surpassed, by several engineers, working independently, in various parts of the world. However, all these early technologists shared one concern in common, "what volume of electricity was needed to power such devices, and how could their performance be measured?"

Fortunately, Edward Weston began producing portable DC measuring equipment around the year 1850, and this radically changed the complexion of early electrical engineering and testing. The concept of the Watt was already well known and had been recognized as a unit of electrical power measurement by the Second Congress of the British Association for the Advancement of Science in 1882. Accordingly, the stage was then set for the optimization of electric motors and generators, but the necessary methodology for achieving same was sorely lacking.

Primarily, this was due to two factors, the identification of all losses associated with electric motor operation had not yet been classified, and the development of a process by which motor efficiencies could be accurately calculated had not yet been solidified.

Initially, it had been assumed that a simple ratio computed by dividing the output power by the input power would provide a reliable yardstick for motor power efficiency. The basic idea was sound, for the closer the quotient came to unity, the nearer the motor efficiency would be to 100%. Hence began the so-called "Quest for Unity."

However, early "electric engines" were very inefficient, probably delivering no more than 30% of the electrical input power to the output shaft. Therefore, it quickly became evident that not all of the applied power was being converted to a mechanical output. Clearly, inefficiencies were involved, and they would have to be accounted for. Until this was complete, the simplified approach to establishing a Factor of Unity could not be realized.

What is remarkable, in the history of motor development, is the curious degree of synergy that seems to have accompanied these early pioneering efforts. Between 1839 and 1850 the British Brew Master James Joule conducted an elegant series of experiments, in which he sought to unify electrical, chemical and thermal phenomena by demonstrating their inter-convertibility and their quantitative equivalence. The results of Joule's work were published in the Philosophical Transactions of the Royal Society, with a very impressive title: "On the Mechanical Equivalence of Heat."

The contributions of Lord Kelvin must be considered next. His paper, "The Dynamical Equivalent of Heat," published in 1851, contended that energy could be "lost to man irrecoverably; but not lost to the material world". Thomson was thus the first person to understand that all energy changes involve energy dissipation, and losses.

During the second half of the nineteenth century Kelvin and other scientists, including Clausius, Rankine, Maxwell and, Boltzmann, continued to develop these ideas. Their combined efforts resulted in the establishment of the Science of Thermodynamics; with Conservation of Energy as its First Law and the Dissipation of Energy as its Second Law.

Accordingly, motor researchers were thus made aware of the extreme importance of classifying all known motor losses, and accounting for them with the same degree of accuracy as would be exhibited in the well-established science of corporate book-keeping.

Ultimately, such practices would lead to the development of two principle kinds of Efficiency Measurements, both very valuable in all research pertaining to rotating components, but, particularly valuable for developing an understanding of over-unity as a scientific fact. Both efficiency concepts shall be fully explained in this presentation.

The Identification of Motor Losses

The modern world is literally overrun with electric motors, so much so, that we tend to take these wonderful machines for granted. Unlike those bygone days of the nineteenth century, there now exist many different types of "electric engines", available in a large assortment of sizes, from hundreds of vendors, including the internet. And, as a result of this "age of electrical plenty," and our isolation from manufacturing and industry in general, few individuals have any concerns about design criterion, manufacturing procedures or motor losses. But, this was not always the case.

Early motor engineers were not only concerned with identifying motor losses, but with measuring them accurately; and this task was not always easy. Originally, the design of a motor prototype, its construction, testing, the cataloging of its losses, and the plotting of its performance curves, was an mammoth undertaking, requiring months of preparation. The undertaking required thousands of man-hours of labor, as there were no computers, calculators or design simulators in those days. Everything was done by hand, and the results were recorded manually in various Engineering Tables.

THE MEANING OF UNITY IN ENERGY CONVERSION SYSTEMS

However, despite the difficulties of that time, motor engineers eventually isolated the following group of losses associated with DC motors and generators:

1.) Iron Losses in the Stator.

2.) Iron Losses in the Rotor.

3.) Copper Losses in Field windings.

4.) Copper losses in the Rotor.

5.) Hysteresis Losses.

6.) Brush and/or Stray Losses.

7.) Frictional Losses.

8.) Windage Losses.

Further research eventually disclosed the fact that these losses exhibited characteristic levels of dissipation depending upon their individual natures, and function within the machine under consideration. For illustration purposes, the following is a modern list of losses derived from the analysis of a 15 Horsepower Induction motor. Said losses are expressed as a percentage of the total input power.

Losses	Watts	Percent	Total Losses	Input Power
Stator	486.522	3.8	1613.204	12,803.20
Rotor	320.08	2.5		
Core	307.277	2.4		
Stray	294.473	2.3		
Friction	136.567	1.066		
Windage	68.284	0.533		

Utilizing the data presented above, we can now determine the overall motor efficiency of the machine in this particular example. This is accomplished with the following relationship:

$$\text{Efficiency} = \frac{\text{Output Power}}{\text{Input Power}} \times 100\%$$

$$\text{Efficiency} = \frac{15 \text{ HP} \times 746 \text{ W}}{12803.204} \times 100\% = \frac{11190.000}{12803.204} = 87.4\%$$

This is a very reasonable efficiency for a modern, high performance motor, for which a customer might be charged a premium price. However, please note that the actual losses tabulated above do not enter directly into this efficiency calculation. Therefore, as a result of using this method of determining overall efficiency, modern engineers never develop the same "feel," as their predecessors possessed, for the numerous relationships that exist among Input Power, Output Power and the assorted losses in any given system.

It is unfortunate, that the modern approach to electrical power analysis has, in many cases, rejected earlier methodologies, simply because they have been deemed obsolete. This researcher believes that such decisions may in fact have been hasty, or perhaps, even erroneous in nature. Fellow investigators deserve to know much more about these forgotten methods, especially now with "Free Energy" being taken seriously in so many circles of endeavor.

Therefore, the following sections shall be designed to shed more light on the methods utilized in the past, and perhaps indicate some unsuspected pathways which may lead to future methodologies, and new discoveries.

Types of Efficiencies and Related Calculations

Effectively, there are two types of efficiencies which shall be presented and explored in this section. The first is System Efficiency (E_s), the second is Conversion Efficiency (E_c). System Efficiency has already been demonstrated above, under the heading of "Overall Efficiency" and is simply the ratio of Output Power to Input Power as previously demonstrated. However, Conversion Efficiency, includes the Loss Parameter in its calculation, thereby supplying an entirely different perspective for the system in question.

E_c = Output Power / (Input Power - Losses)

A meaningful comprehension of this relationship can be gained by means of an illustrative example. Note, that multiplication by 100% has been left out of the above formula. This has been done deliberately and is intended to make the point of this exercise that much more obvious. However, it must be borne in mind, that motor losses were identified and accurately measured over a long period of time, such information was not always available in its entirety as it is today.

THE MEANING OF UNITY IN ENERGY CONVERSION SYSTEMS

Consider, now, the list of losses presented above, in association with the 15 HP induction motor. For the sake of this demonstration, we shall assume that each loss parameter was identified and measured in the same order that they appear above, and thereafter used to compute a revised Conversion Efficiency. In so doing, we shall develop six separate calculations, and thereby call attention to a trend that is usually ignored completely.

1.) Stat : E_c = 11190 / (12803.204 - 486.522) = .9085

2.) Stat+Rot : E_c = 11190 / (12803.204 - 806.602) = .9327

3.) Stat+Rot+Cor : E_c = 11190 / (12803.204 - 1113.879) = .9573

4.) Stat+Rot+Cor+Stra : E_c = 11190 / (12803.204 - 1408.752) = .9821

5.) Stator+Rotor+Cor+Stra+Frict : E_c = 11190 / (12803.204 - 1545.319) = .9939

6.) Stat+Rot+Cor+Stra+Frict+Windage : E_c = 11190 / (12803.204 - 1613.204) = 1.0000

Please note, that each addition of a loss segment moves the Conversion Efficiency toward the value of 1.000, or **Unity**! Thus, during the early days of motor development, "the pursuit of unity" was considered a vital undertaking, for a Conversion Efficiency of Unity was originally considered a reliable indication that all losses in a given system had been properly accounted for.

However, the arrival and implementation of Conservation of energy principles, soon made this practice obsolete. For anyone suitably familiar with the laws of Physics, would automatically realize that "Unity" was indeed the upper limit of an energy accounting process, which could never yield a value greater than one, unless a mistake had been made. Nonetheless, the basic procedure is still in use today, and is now known as a **Segregated Load Analysis.** The modern interpretation of these procedures can be found in IEEE 112-B; International Standards for Induction Motor Efficiency Evaluation. However, while it is fine to be informed, don't be seduced by these modern shortcuts, for there is much more to follow.

Merging Two Efficiencies

The use of System Efficiency is a universal practice in many scientific disciplines, and is, therefore, well known and completely understood. Conversion efficiency represents a process which is believed to be almost useless in the modern world, and aside from those few who may still teach it for the sake of clarification, it has been discarded by the scientific and academic communities, in favor of the teachings of conservation of energy. Accordingly, one might rightfully ask the question, what purpose does this outdated calculation represent?

The best way to explain the usefulness of Conversion Efficiency is to study its effects upon System Efficiency, and there are two ways to do this. Consider the following expressions:

$$\text{Output Power} = E_s \, (\text{Input Power})$$

and

$$\text{Output Power} = E_c \, (\text{Input Power} - \text{Losses})$$

THE MEANING OF UNITY IN ENERGY CONVERSION SYSTEMS

Mathematically, both these statements are equivalent, and so they may be equated one to the other. Such an arrangement yields the following relationship:

$$E_s \text{ (Input Power)} = E_c \text{ (Input Power - Losses)}$$

Now, solving for the System Efficiency, we obtain:

$$E_s = E_c \frac{\text{(Input Power - Losses)}}{\text{(Input Power)}} = E_c \frac{\text{(Output Power)}}{\text{(Input Power)}}$$

This quotient expresses the System Efficiency as a decimal relationship, involving the Input Power, the Losses, and the Conversion Efficiency. Multiplication by 100% will express the quotient as a percentage, which is the usual format expected for an efficiency rating.

More importantly, however, we can now appreciate the System Efficiency from an entirely different perspective, which happens to involve the Conversion Efficiency and the Input Power. Naturally, in all standard systems, E_c is equal to one, and so it may be left out if E_s is being calculated for a system adhering to classical limitations.

The second method of deriving a relationship between System Efficiency and Conversion Efficiency, begins by solving each of the initial equations for the Input Power. Accordingly,

THE MEANING OF UNITY IN ENERGY CONVERSION SYSTEMS

$$\text{Input Power} = \frac{(\text{Output Power})}{E_s}$$

and

$$\text{Input Power} = \frac{(\text{Output Power} + \text{Losses})}{E_c}$$

Again, because both these statements are equivalent, they may be equated one to the other. Such an arrangement yields the following relationship:

$$E_s = \frac{E_c \ (\text{Output Power})}{(\text{Output Power} + \text{Losses})}$$

As in the previous solution, this quotient expresses the System Efficiency as a decimal relationship, involving the Output Power, the Losses, and the Conversion Efficiency. Multiplication by 100% will express the quotient as a percentage, which is the usual format expected for an efficiency rating.

Additionally, we can now appreciate the System Efficiency from an entirely different perspective, which now involves the Conversion Efficiency and the Output Power. Naturally, in all standard systems, E_c is again equal to one, and so it may be left out if E_s is being calculated for a system adhering to classical limitations.

The validity of both these solutions can easily be proven by substituting values of Input Power, or Output Power, from the previous example, and computing the System Efficiency in each case. The resulting motor performance should be 87.4% in each case, as long as Ec = one.

The reader may be wondering, at this point, what is the purpose of maintaining the Ec parameter, in these calculations, if it must always have a value of one? The best way to expound upon this point, is by means of an example. However, two points should be remembered:

a.) The value of "Unity," is only particular to classical, linear, electromagnetic systems.

b.) The actual function of the Ec parameter may be very different from what is presently assumed. This point deserves further exploration.

Having made these statements, let us return to the example of the 15 HP Induction Slip Motor presented in the first part of this report, under the heading of The Identification of Motor Losses.

For the sake of this example, we shall assume that measurements have just been made upon this device for the first time. We discover that all the losses are exactly the same as previously tabulated, the Input Power is consistent at **12803.204** electrical watts, but suppose the Output Power is measured at **14684.97522** shaft watts. Because the intention here is to determine the System Efficiency, we will start off by utilizing the formula for Es in standard format.

$$Es = \frac{Ec \ (\text{Output Power})}{(\text{Input Power})} \times 100\%$$

THE MEANING OF UNITY IN ENERGY CONVERSION SYSTEMS

Here, naturally, the compulsion will exist to assume that E_c must be unity, therefore, the initial calculation will most probably yield the following result:

$$E_s = \frac{(14648.97522)}{(12803.204)} \times 100\%$$

$$E_s = 114.416479\%$$

Admittedly, this result is totally hypothetical, but if it actually happened, it would take the conditioned brain a few seconds to realize the implications of such a result, and once the notion of "Impossibility" had registered, the most natural tendency would be to search for a mistake. Therefore, the measurements and the figures would be checked and re- checked, but if no errors were discovered, what then? Everything seems to check out, yet, something **must** be wrong. The Engineer stares at the data in empty-headed fashion; doubting, hoping, wondering!

Then, a sudden realization bursts upon the exhausted mind; the actual losses were not included in these calculations, surely an error must be reside with the losses somehow. A quick study of the available formulas implies that the data can be used in reverse to check system losses by using the relationship E_s **(Input Power) =** E_c **(Input Power - Losses)**, and solving for the Losses at hand. Accordingly, you reconfigure for the equation, and re-substitute your values.

THE MEANING OF UNITY IN ENERGY CONVERSION SYSTEMS

Hence:

Losses = Input Power - [(Es / Ec) (Input Power)]

Losses = (12803.204) - [(1.14416479) / (1) (12803.204)]

Losses = - 1845.771216 watts

However, this answer makes no sense whatsoever! Not only is the numerical value incorrect, but so is the polarity! Therefore, something else must be wrong. The only thing left to consider at this point, is the belief that Ec must be equal to one. Perhaps that supposition was incorrect from the start. Judging from the structure of the original equation, either Es or Ec could have been made equal to unity. This situation demands further experimentation! Therefore, swap Es & Ec:

Losses = Input Power - [(Es / Ec) (Input Power)]

Losses = (12803.204) - [1 / (1.14416479) (12803.204)]

Losses = (12803.204) - [(.8739999769) (12803.204)]

Losses = (12803.204) - (11190)

Losses = 1613.204

Now the answer is correct, which means that the Conversion Efficiency could never have had a value of one, as demonstrated above. But, of equal importance, is the fact that the System Efficiency now suggests a value of 100%! So, by these simple calculations, a most important feature of this investigation is propelled into the foreground; how can any system be 100% efficient, and still support losses of 1615.204 watts? Is there any way to understand this apparent dilemma? The answer is yes.

But, before an explanation can be expounded upon, it will first be necessary to review some very fundamental concepts which are particular to all motors and generators, but which are not necessarily understood from the most advantageous point of view!

Force, Voltage and Power in Rotating Components

The technological history of motors and generators is a long and convoluted story, which has resulted in the emergence of power components that spin within a magnetic field. Accordingly, all such machines have certain features in common, and these common elements automatically determine the overall features characteristic of these rotating machines.

Almost all textbooks that teach modern electrodynamics will have sections devoted to the motional EMF generated by a conductor moving in a magnetic field, and a section devoted to the production of a force upon a current-carrying conductor immersed within a magnetic field. These electromagnetic effects come under the headings of Faraday's motionaly induced EMF and applications of Lenz Law respectively.

Mathematically speaking, the induced voltage can be related to three parameters, which are very familiar to most physics students:

$$V = B \ell v$$

where, V is the induced voltage, B is the magnetic flux density, ℓ is the length of the conductor, and v is the velocity of that the conductor within the magnetic field.

On the other hand, the force acting upon a current-carrying conductor can be described by the following relationship:

$$F = B \ell I$$

where, F is the force, B is the magnetic flux density, ℓ is the length of the conductor, and I is the current moving through the conductor.

For the most part, these basic relationships are so familiar to most engineers, that they are literally not given a second thought. It is a well-known fact that these expressions are basically correct, and fundamentally accurate, so what else is there to consider?

This researcher has made it a general practice to consider every essential process from the standpoint of the power involved. For only with the consideration of actual power can we develop a holistic "feel" for a given system, and the manner in which it will perform.

Voltage is only a single component of electrical power. The entire picture requires the inclusion of an electric current in order to become completely meaningful. Therefore, to produce an expression in electrical power, the above equation for voltage must be multiplied by current (I), yielding:

$$VI = B \ell v I$$

THE MEANING OF UNITY IN ENERGY CONVERSION SYSTEMS

Similarly, Force is only one component of mechanical power. The understanding of the entire picture requires the inclusion of a velocity to become completely meaningful. Therefore, to produce an expression in mechanical power, the above equation for force must be multiplied by a velocity (v), yielding:

$$Fv = B \ell I v$$

Thus, we immediately see the benefit of expressing both of these fundamental relationships in terms of their powers, for suddenly, they become equal entities, rather than unrelated expressions. Accordingly, this process has altered the consciousness of the observer, for now it is not only evident that an equality exists, but it is evident why it exists. However, there is more to consider.

Because **B ℓ v I** is equal to **B ℓ I v**, it is evident that **VI** must be equal to **Fv**, but what does this actually mean? Certainly, it suggests that the electrical power is equal to the mechanical power, but actually, it means quite a bit more than that. For in an actual electrical machine, be it motor or generator, B, **ℓ, I** and **v** must all exist simultaneously during operation! Therefore, all motors are generators, and all generators are motors, at exactly the same time. The only difference between the two situations resides in the direction of power flow, and in what is being "generated".

Considering the case of the electric motor, the power is flowing from the electrical input to the mechanical output. Thus, positive mechanical power is being delivered to the output shaft. However, the machine is also "generating" a voltage as it spins. This voltage represents a potential which is directed in opposition to the voltage supplied by the source. Accordingly, the driving current must flow against the generated Back EMF, thereby developing negative power at the input terminal and causing a condition referred to as Degradation of Electrical Potential.

Conversely, if the generator case is considered, the power is flowing from the mechanical input to the electrical output. Thus, positive electrical power is being delivered to the generator's load terminals. But, in this case, the machine is also "generating" a torque as it spins. This torque represents a torsional effort which is directed in opposition to the torque supplied by the prime mover. Accordingly, the driving torque must operate against the generated Back Torque, thereby developing negative power at the input shaft and causing a condition referred to as *Degradation of Mechanical Potential*.

Normally, these manifestations of mechanical and electrical power are encouraged to seek a condition of equilibrium, between electrical power supply and electrical motor forces plus mechanical load. Or between the mechanical power supply and the generated mechanical forces plus electrical load. Under the conditions so defined, the entire system achieves an equilibrium based on a complete cancellation of the powers involved but can never reach an equilibrium with respect to the consumables supplied by the source; namely, the current or the torque.

However, studying such systems from the unusual point of view herein presented, now allows an exploration of another set of conditions, not usually considered, especially in so-called linear electromagnetic systems. It is a condition in which an apparently normal electro-mechanical system can be prompted into a mode of operation, which causes disequilibrium to occur for a portion of each operational cycle. Thereby disallowing a total cancellation of the powers involved, but effectively encouraging a totally new kind of steady-state condition which manifests within the applied mechanical torque, or the consumed electrical current.

The Dynaflux Alternative

Thirty years ago, on October 25, 1988, United States Patent Number 4,780,632A, entitled *"Alternator With Improved Efficiency,"* was issued to this researcher. A brief summary of the invention, extracted from the actual patent text, will serve to acquaint the reader with this technology, should he be unaware of its existence.

It is therefore an object of the present invention to provide an alternator operating at significantly higher efficiencies than those known to the art.

A further object of this invention is to provide an alternator capable of producing a higher-frequency output per shaft revolution per pole set than is known to the art.

Another object of this invention is to provide an alternator which can derive useful work from the reactive power which usually contributes only to the losses in present alternating current generating systems.

These and other objects are achieved in the present invention by providing an alternator with a stator element carrying a combination of field windings and power windings and a rotor which reciprocates the magnetic field across said power windings with substantially simple harmonic motion.

One embodiment of the invention employs a rotor fabricated from a stack of laminated disks, pressed upon an arbor which is obliquely disposed with respect to the intended axis of rotation, and integrally machined in order to provide the assembly with a peripheral contour equivalent to that of a cylinder.

The stator is formed of two salient pole projections, each having a concave pole face whose radius is slightly greater than the radius of the rotor. The rotor thereby defines an air gap of continuous dimension when rotated. The rotor is in series with the two pole pieces to complete the magnetic circuit.

As the rotor is rotated, the zone at which the flux is coupled to the pole pieces varies in position along the length of each pole face. The magnetic flux is swept forward and then backward across the power windings during each revolution, thereby inducing a voltage therein.

Since the magnetic flux reverses its direction of motion, but never its intrinsic polarity, there is a generation of force at the point of reversal which acts in concert with the force provided by the prime mover. Thus, an alternator constructed according to the present invention will produce a given volume of output power for a significantly reduced volume of input power.

Altogether, it took nearly four years for this patent to finally make its way through the legal labyrinth in Washington, D.C., and eventually attain the Patent Granted status that it so honestly deserved. Most of this delay, was due to the steadfast refusal of one stubborn patent examiner, his name shall go unmentioned in this document, who insisted that the technology should be rejected, on the grounds that it implied a violation of physical law.

This unwarranted conclusion was based, in part, upon the last paragraph cited above, and upon the accompanying data, which was disclosed within the patent application, as a means of supporting this very controversial pronouncement.

Ironically, the invention was never accused directly of violating conservation of energy, it was simply rejected, time and time again, for purely nonsensical reasons, until the examiners ran out of arguments. And, there are logical reasons for developing these indirect attacks upon technologies undesired by any governmental system.

Their strategy is to force the applicant to focus upon the exact nature of the rejection, no matter how ridiculous, rather than allow him the latitude of resorting to a more natural and universal defense possibly available directly from Nature itself.

But here, the patent in question has already been issued. The accompanying information has been residing in the public domain for thirty years. Therefore, this writer, is no longer constrained by threats of rejection from the boys in Washington and shall now resort to explaining the merits of the Dynaflux Alternator by resorting to a fundamental example which relies entirely upon existing, classical laws of Physics.

THE MEANING OF UNITY IN ENERGY CONVERSION SYSTEMS

The objective here is to explain an unexpected increase in Kinetic Energy, which might be commercialized upon, if the properly designed equipment existed, with which to intercept the gain developed. To elucidate this procedure, we shall make use of an impulse device, such as a military mortar, so adjusted as to launch a projectile in an absolutely vertical direction. For the purposes of this example, such outside forces as air friction and wind velocity shall be assumed as non-existent. Case I, shall be completely restorative. Case II shall not be restorative.

Consider a ten-pound, spherical projectile, hereinafter referred to as the "sphere." The mass of this sphere, in English Units, shall be .3125 slugs. The driving impulse (I_M), is defined as force times time, or slug-ft / second. Therefore:

Mass (m) = .3125 slugs.
Force (F) = 500 pounds.
Time (t) = .1 seconds.
Impulse (I_M) = 50 slug-ft / sec.
Gravitational Acceleration (g) = 32 ft / sec².

Therefore:

$$I_M = Ft = mv, \&, v = (Ft) / m.$$

Case I: Sphere rising by Impulse & falling under the influence of gravity: Restorative.

a.) A force of 500 pounds acts on a .3125 slug sphere for .1 seconds.

$$I_M = Ft = 50 \text{ slug-ft / sec.}$$

THE MEANING OF UNITY IN ENERGY CONVERSION SYSTEMS

b.) Find the resulting Velocity:

$$v = (Ft)/m = [(50 \text{ slug-ft/sec})/(.3125 \text{ slugs})] = 160 \text{ feet/second}$$

c.) Find the time of the sphere rising:

$$V_0 - gt = 0$$
$$160 - 32t = 0$$
$$-32t = -160$$
$$t = 5 \text{ seconds}$$

d.) Find the height attained by the sphere:

$$h = v_0 t - \tfrac{1}{2} g t^2$$
$$h = (160 \times 5) - (16 \times 25)$$
$$h = (800) - (400)$$
$$h = 400 \text{ feet}$$

e.) Calculate the Potential Energy at maximum Height:

$$PE = mgh$$
$$PE = (.3125 \times 32 \times 400)$$
$$PE = 4000 \text{ ft-lbs}$$

f.) Calculate the Kinetic Energy when Sphere returns to ground level.

$$KE = \tfrac{1}{2} m v^2$$
$$KE = (\,.5 \times .3125 \times 160^2\,)$$
$$KE = 4000 \text{ ft-lbs}$$

The Kinetic Energy attained at ground level is exactly equal to the Potential Energy at 400 feet of elevation. And, had it been calculated, the same value would be applicable to the initial velocity which propelled the sphere away from the earth. Thus, the overall performance can be characterized as energy conservative, which should have been expected, considering the fact that gravity is a restorative force. However, in the following example, certain modifications shall be employed, which shall drastically alter the results, and perhaps supply some mind-opening realizations concerning the nature of the Dynaflux Alternator!

Case II: Sphere rising by Impulse & falling under impulse + gravity: Non-Restorative.

a.) A force of 500 pounds acts on a .3125 slug sphere for .1 seconds.

$$I_M = Ft = 50 \text{ slug-ft / sec.}$$

b.) Find the resulting Velocity:

$$v = (Ft)/m = [\,(50 \text{ slug-ft / sec})/(.3125 \text{ slugs})\,] = 160 \text{ feet / second.}$$

THE MEANING OF UNITY IN ENERGY CONVERSION SYSTEMS

c.) Find the time of the sphere rising:

$$V_0 - gt = 0$$
$$160 - 32t = 0$$
$$-32t = -160$$
$$t = 5 \text{ seconds.}$$

d.) Find the height attained by the sphere:

$$h = v_0 t - \tfrac{1}{2} g t^2$$
$$h = (160 \times 5) - (16 \times 25)$$
$$h = (800) - (400)$$
$$h = 400 \text{ feet}$$

e.) Calculate the Potential Energy at maximum Height:

$$PE = mgh$$
$$PE = (.3125 \times 32 \times 400)$$
$$PE = 4000 \text{ ft-lbs}$$

f.) Add a second impulse in the downward direction, and calculate time of descent:

$$h = v_0 t + \tfrac{1}{2} g t^2$$

$$400 = 160t + 16t^2$$

$$25 = 10t + t^2$$

$$0 = t^2 + 10t - 25$$

THE MEANING OF UNITY IN ENERGY CONVERSION SYSTEMS

Solving for t with the quadratic formula, we obtain:

t = 2.07 seconds, the time of descent

g.) Calculate the final Velocity at ground level:

$$V_F = V_0 + gt$$

$$V_F = 160 + 32t$$

$$V_F = 160 + 66.24$$

$$V_F = 226.24 \text{ feet / second}$$

h.) Calculate the final Kinetic Energy at ground level:

$$KE_F = \tfrac{1}{2} mv^2$$

$$KE_F = (\tfrac{1}{2})(.3125)(226.24^2)$$

$$KE_F = (\tfrac{1}{2})(.3125)(51184.5376)$$

$$KE_F = 7997.584 \text{ ft-lbs}$$

i.) Calculate change in Kinetic Energy:

$$\Delta KE = KE_F - KE$$

$$\Delta KE = (7997.584) - (4000)$$

$$\Delta KE = 3997.584 \text{ ft-lbs.}$$

THE MEANING OF UNITY IN ENERGY CONVERSION SYSTEMS

The Kinetic Energy in this example is nearly doubled. However, the critics shall immediately have a negative comment, such as "So what, you added a second driving impulse, equal to the first!" This observation is exactly correct.

But, bear in mind, that if the demonstration could be performed on a horizontal plane, the Kinetic Gain would vanish completely, and all that would remain would be a constant velocity oscillator, producing 4000 ft-lbs of energy on every half cycle. Thus, it is the gravitational force component that is responsible for the asymmetry in the system's Kinetic Energy. And, the reason for the importance of this fact, resides in the realization that **the gravitational force is free!**

A similar set of conditions prevail in the Dynaflux Alternator. A mechanically driven rotor provides impulse at both limits of the generator's magnetic circuit, while the interaction between the generated current and the field flux, provides the resisting force, which is analogous to the force of gravity. **This force is called the Lenz Force, and, just as in the case of gravity, it is basically unavoidable, so why not take advantage of it!**

These facts have been sanctioned by the United States Government in the release of US Patent # 4,780,632A! **A Power Accounting of the Dynaflux Alternator.**

The original Dynaflux Alternator prototype was a very primitive, home-made device, quickly constructed from assorted bits of junk that were laying around. It was not laminated, the steel was not of a high magnetic grade, and the power coils were probably over-wound. In retrospect, the machine should not have been patented so early in its development. However, due to the haste of youth, caution was thrown to the winds.

All power measurements were made with ancient electrodynamic watt meters, but a segregated load analysis was performed just the same, the results of which appear below.

Losses	Watts	Percent	Total Losses	Input Power	Output Power
Field	145.950	27.839	510.400	524.246	036
Iron	219.870	41.940			
Stray	034.936	06.664			
Friction	071.227	13.587			
Windage	038.417	07.328			

Not only were the results pathetic, but the numbers did not make sense no matter how they were interpreted. Initially, this researcher had no realization concerning the existence of Conversion Efficiency. However, an article concerning the same was eventually located in a college library archive, and with that information, the mystery was eventually solved.

The Dynaflux Conversion Efficiency computed as follows:

E_c = Output Power / (Input Power - Losses)

$$E_c = \frac{36}{(524.246 - 510.400)} = \frac{36}{13.846}$$

$E_c = 2.6000$

Such a number seemed quite amazing, yet completely unrealistic, for the System Efficiency was nearly non-existent!

$$E_s = \text{Output Power} / (\text{Input Power}) \times 100\%$$

$$E_c = \frac{36}{524.246} = 6.6867\%$$

Certainly not a very impressive result. But, after having combined E_s and E_c into one expression, things suddenly began to make sense!

While working with the expression for the input Power in terms of E_c, the Output, and the Losses, something very remarkable suddenly became apparent to me.

$$\text{Input Power} = \left(\frac{\text{Output Power}}{E_c}\right) + \text{Losses}$$

At that time, the relationship between the Input Power, and E_c was still most confusing, and the actual Dynaflux Data was being studied in all manner of ways in an attempt to shed light on the overall situation. Most importantly, it was necessary to investigate the effect of such a large value of E_c upon the Input Power. Therefore, actual values were substituted into the above expression.

THE MEANING OF UNITY IN ENERGY CONVERSION SYSTEMS

$$\text{Input Power} = \frac{(36.000)}{2.6} + (510.40)$$

$$\text{Input Power} = (13.8461) + (510.40)$$

$$\text{Input Power} = 524.2461 \text{ watts.}$$

This result was gratifying, because it validated both the calculations, and the measurements. However, it immediately raised the question, "What then would result from a Conversion Efficiency of Unity?" Accordingly. The calculation was redone with E_c equal to 1.

$$\text{Input Power} = (36.000) + (510.40)$$

$$\text{Input Power} = 546.40 \text{ watts}$$

A considerable larger value! But what did this really mean? The Output Power was rock-stable at 36 watts, the Input Power was quite constant at 524.246 watts, and the E_c had been determined using these values plus the measured losses. So, it was reasonable to assume that the E_c might also affect the Losses in some way. However, calculating the losses in both the above cases, gave two very different numbers! Allowing E_c to equal 2.6, suggested ridiculously high losses of 1327.03986 watts, conversely allowing E_c to equal 1, produced the more reasonable result of 488.2461 watts. But, what did these numbers actually mean?

Finally, a thought occurred, what is represented by the difference between the 546.400 watt input, and the 524.2461 watt input? **A simple subtraction produced a figure for Δw of 22.1539 watts.** Initially, this figure meant nothing at all.

THE MEANING OF UNITY IN ENERGY CONVERSION SYSTEMS

However, some deep reasoning finally served to put the matter into proper perspective.

If more Input Power was required when E_c was unity, than was required when E_c was valued at 2.6, then there had to be an additional input to the system supporting this condition!

We already know of a wattage differential of magnitude 22.1539, which was generated by subtraction, on the left hand side of the Input Power equation. What would be the result of transposing this value to the right-hand side? Accordingly:

Losses = 546.40 w - 36 w - 22.1539

Whereupon, Losses = 488.246 watts, not 510.40 watts!

But, proceeding further, we discover the ultimate truth:

$$510.40 \text{ w} + \chi = 488.2461 \text{ w, and}$$

$$\chi = 488.2461 \text{w} - 510.40 \text{ w}$$

$$\chi = -22.1539 \text{ watts}$$

Recall, that this exact number has made an appearance previously, but without the negative sign! **This change of polarity is most significant, for it indicates a negative loss, which is actually an Assistive Power Component!** Herein, then, lies the basic nature of the Dynaflux Generator; a machine which has the ability to supply an assisting force to the prime mover, at very specific points in each cycle, **by becoming a motor!** Accordingly, the Dynaflux Losses consist of two distinct components, a positive value, and a negative value. The importance of this novel expression for the Total Loss requires that it be expressed as an algebraic sum, for the sake of absolute clarity. As well as, a distinct and unique symbol, possibly (L_D), thus:

Total Dynaflux Losses (L_D) = [Actual Losses + Negative Losses]

$$(L_D) = [\ 510.40 \text{ w} + (\ -22.1539 \text{ w}\ )\ ]$$

$$(L_D) = [\ 488.2461 \text{ w}\ ]$$

Finally, now, the Dynaflux Input Power relationship can be written clearly, with its full implication mathematically displayed:

Input Power = Output Power + Losses

$$(\ 546.40 \text{ w} - 22.1539 \text{ w}\ ) = 36 \text{ w} + [\ 510.40 \text{ w} + (\ -22.1539 \text{ w}\ )\ ]$$

$$(\ 524.2461 \text{ w}\ ) = 36 \text{ w} + [\ 488.2461 \text{ w}\ ]$$

Most of the disclosures made within this section of my report will, no doubt, be very exciting to those readers possessing a genuine interest in this subject matter. However, this section would not be complete without presenting one final and very important fact. For anyone studying **The Meaning of Unity**, for the sake of pursuing the concept of Over-Unity, the following teaching should prove to be invaluable. This researcher is now referring to the quantity of Power Assisting Loss, or PAL, which is the acronym used to describe Negative Loss, the parameter previously defined and derived. We shall start with the following relationship:

$P_{AL} = [\,(\text{Output Power} / E_C) + \text{Losses} \,] - [\,\text{Output Power} + \text{Losses} \,]$

$P_{AL} = 1/E_C [\,\text{Output Power} + E_C (\text{Losses}) \,] - [\,\text{Output Power} + \text{Losses} \,]$

$P_{AL} + [\,\text{Output Power} + \text{Losses} \,] = 1/E_C [\,\text{Output Power} + E_C (\text{Losses}) \,]$

$E_C [\,P_{AL} + \text{Output Power} + \text{Losses} \,] = [\,\text{Output Power} + E_C (\text{Losses}) \,]$

$[\,E_C P_{AL} + E_C \text{Output Power} + E_C \text{Losses} \,] = [\,\text{Output Power} + E_C (\text{Losses}) \,]$

$[\,E_C P_{AL} + E_C \text{Output Power} + E_C \text{Losses} - E_C \text{Losses} \,] = [\,\text{Output Power} \,]$

$[\,E_C P_{AL} + E_C \text{Output Power} \,] = [\,\text{Output Power} \,]$

$E_C [\,P_{AL} + \text{Output Power} \,] = [\,\text{Output Power} \,]$

$[\,P_{AL} + \text{Output Power} \,] = [\,\text{Output Power} \,] / E_C$

$P_{AL} = [\,\text{Output Power} / E_C \,] - [\,\text{Output Power} \,]$

Or, in true abbreviated form:

$$P_{AL} = (P_{OUT} / E_C) - P_{OUT}$$

Thus, it is interesting to note, that all loss-related parameters cancel out, leaving only the Output Wattage and the Conversion Efficiency to deal with. However, with it, one can easily obtain the Power Assisting Loss, or the Negative Loss.

THE MEANING OF UNITY IN ENERGY CONVERSION SYSTEMS

Increasing Dynaflux Conversion Efficiency

The original Dynaflux Alternator prototype had only one redeeming quality, and that was its Conversion Efficiency. That tiny machine loaded with a mere 36 watts, provided more insight into Tesla's secrets, than any huge Poly-phase generating Goliath could possibly have provided. It is most important that the meaning of this Conversion Efficiency Effect be completely understood, before we proceed to other important concepts. Therefore, let us explore the E_C Factor, one point at a time.

First of all, using the original Dynaflux Alternator Data, let us re-calculate the standard System Efficiency, deliberately excluding the Conversion Efficiency Factor completely.

$$E_S = \frac{(\text{Output Power})}{(\text{Input Power})} \times 100\%$$

$$E_S = \frac{36}{524.2461} \times 100\% = 6.867\%$$

Not exactly the kind of result that would prompt an investor to part with millions of dollars! But, let us now examine what happens to the exact same data when the effects of the Conversion

THE MEANING OF UNITY IN ENERGY CONVERSION SYSTEMS

Efficiency are considered. To do this, we shall utilize the System Efficiency first displayed on page 17 and 18 of this book:

$$E_s = \frac{E_c \text{ (Output Power)}}{\text{(Input Power)}} \times 100\%$$

$$E_s = \frac{2.6 \times 36}{524.2461} \times 100\% = 17.854\%$$

The Efficiency increase between 6.867% and 17.854%, represents an improvement of 2.6 to one, exactly the same magnitude as the Conversion Efficiency, which leaves no doubt that this parameter is producing the overall effect. But, just how far can this concept be pushed, and what would be the ramifications of such actions if they could be commercially implemented?

In order to investigate this situation, this writer has opted to utilize the exact same losses originally measured in the first Dynaflux prototype, but, allowing for an enhanced power output, and a greatly increased Conversion Efficiency. The purpose of this exercise is not to present any realistic engineering information, but rather, to indicate by example, the manner in which such accomplishments may be achieved.

It would be useful to realize, that certain Natural limitations to this process may actually prevail in the real world. But, such limits notwithstanding, the intended message shall be clear.

THE MEANING OF UNITY IN ENERGY CONVERSION SYSTEMS

Consider now, the following data:

Losses	Watts	Percent	Total Losses	Input Power	Output Power
Field	145.950	27.839	510.400	524.246	501.2288
Iron	219.870	41.940			
Stray	034.936	06.664			
Friction	071.227	13.587			
Windage	038.417	07.328			

Resorting to the standard method of calculating the **System Efficiency**, we find the following:

$$Es = \frac{(\text{Output Power})}{(\text{Input Power})} \times 100\%$$

$$Es = \frac{501.2288}{524.2461} \times 100\% = 95.600\%$$

This is certainly a drastic improvement over the performance of the original Dynaflux Machine. It will be most important to understand what factors have produced such incredible changes. But, in order to understand this result completely, the facts must be presented in four distinct steps:

Step # 1:

Utilizing a Conversion Efficiency of 36.2 and an output of 501.2288 watts, calculate the system Input Power. This parameter shall be identified as Input 1.

THE MEANING OF UNITY IN ENERGY CONVERSION SYSTEMS

Input 1 = $\dfrac{\text{Output Power}}{E_C}$ + Losses } Where E_C = 36.2

Input 1 = $\dfrac{501.2288}{36.2}$ + 510.400

Input 1 = 524.2461 watts

The reader must understand that this is strictly a manufactured result, predetermined to make a point. In no way does it represent actual data derived from Dynaflux, or any other operational device.

Step # 2:

Utilizing an output of 501.2288 watts, and losses in the amount of 510.400 watts, recalculate the system Input Power utilizing a Conversion Efficiency of Unity. This parameter shall be identified as Input 2.

Input 2 = $\dfrac{\text{Output Power}}{E_C}$ + Losses } Where E_C = 1.000

Input 2 = $\dfrac{501.2288}{1.000}$ + 510.400

Input 2 = 1011.6288 watts

The purpose of making this computation from the perspective of Unity Conversion Efficiency, is to establish a relativistic

THE MEANING OF UNITY IN ENERGY CONVERSION SYSTEMS

point of view with respect to the universally accepted method of determining the total Input Power.

Step # 3:

Calculate the Differential Input Power, by subtracting Input 2 from Input 1.

Δ Input = Input 1 - Input 2

Δ Input = (524.2461 - 1011.6288)

Δ Input = - 487.3827 watts (Negative Losses)

Notice that this difference is represented by a negative number! Notice also, that the same result may be acquired by using the Power Assisting Loss relationship.

Step # 4:

Recall, now the relationship:

Input Power = Output Power + Losses } Replace Losses with Total Losses, where
Total Losses = (Losses + Negative Losses), ∴

Input Power = Output Power + Total Losses

Input Power = Output Power + (Losses + Negative Losses)

Accordingly:

Input Power = Output Power + Losses } Replace Losses with Total Losses, where
1011.6288 =   501.2288   + 510.4   Total Losses = (Losses + Negative Losses)
Total Losses = 510.4 + (- 487.3827 watts)

THE MEANING OF UNITY IN ENERGY CONVERSION SYSTEMS

However, the same value must be added to both sides of the equation, to maintain balance, ∴

$$1011.6288 + (-487.3827) = 501.2288 + [\,510.4 + (-487.3827)\,]$$

$$1011.6288 + (-487.3827) = 501.2288 + [\,23.0173\,]$$

} Note that the Negative Power Component consumes all but 23.0173 watts of Loss.

And Finally,

524.2461 watts = 524.2461 watts

Thus, it can now be seen, that the "free resisting force" provided by Lenz Law can be utilized in a machine like Dynaflux, for the purpose of "consuming" its own losses, and thereby boosting the System Efficiency Considerably!

In this exaggerated example, a pathetic little machine like my first prototype, operating at an efficiency (E_s) of only **6.867%**, enjoyed a boost in performance all the way up to **95.600%** by making use of electro-mechanical feedback, and the Lenz Law, which can always be obtained for free! The psychological difference lies in the fact that the Lenz Reaction, like all laws of Nature, can either be worked with, or worked against, and the outcomes will differ as night and day!

Adhering to Tesla's Advice

One hundred and thirty years ago, Tesla delivered a series of lectures, in which he gave some very worthwhile advice concerning the methods he used for designing and building his electro-mechanical prototypes. When I first read those suggestions, I took them to heart simply because they were spoken by the master himself. However, it was not until many years later, that I actually began to appreciate the wisdom inherent in Tesla's words.

One of his most profound statements has guided me in my personal engineering practices over the last fifty years, and from these same words, I will introduce a series of rules which, if followed, will save the independent researcher untold months of frustration, confusion, and lost time. So, without further delay, here is Nikola Tesla's Cardinal Rule:

THE MEANING OF UNITY IN ENERGY CONVERSION SYSTEMS

"Design and build all your machines as close to 100% efficiency as humanly possible!"

Rules to Assist the Independent Investigator in Over Unity Research.

1) Attempt to build all prototypes as close to 100% efficiency as possible.

2) Do not take short-cuts with power measuring equipment, strive for high accuracy.

3) Take multiple readings on each parameter and average the results.

4) Generate a comprehensive Segregated Load Analysis, as soon as possible.

5) Look to National Bureau of Standards, and IEEE to identify all losses involved.

6) Determine your System Efficiency (E_s) early-on, strive for high numbers.

7) Determine your Conversion Efficiency (E_c) if you are well above Unity, Great!

8) If Over Unity, Investigate your data in the Merged Efficiency relationships.

9) If Over Unity, Calculate your Power Assisting Loss Factor or Negative Losses.

10) Add your PAL to your Total Losses and note the magnitude of the remainder.

Remember, that an over Unity condition can hide in a device displaying a low System Efficiency!

Conclusion

There are at least four recognized levels of Entropy Production associated with all Energy Conversion Technologies, and each may be categorized by studying the Conversion Efficiency and the manner in which the Negative Losses interact with the Segregated Losses and the Input Power within a given system.

Positive Entropy System: Standard, linear systems, in which the losses stand alone, and are not affected by any feedback mechanism, in other words, E_c is Unity.

Positive Entropy Diminished: A system in which the losses are diminished by the introduction of Negative Losses but are not completely consumed; $E_c > 1$.

Zero Entropy System: A system in which the losses are completely consumed by the development of Negative Losses, such that E_c is a large Positive number.

Negative Entropy System: A system in which the Negative Loss production entirely consumes all the losses, and most, or all of the Input Power.

Admittedly, these "Rules" represent an introduction to a paradigm which is completely in its infancy. However, this researcher believes that a lifetime of independent research can only be deemed useful if it is shared with other researchers and utilized as a nucleus from which the growth and propagation of knowledge can occur. Therefore, I wish to encourage all my fellow discoverers to continue pushing forward in this field of **Over Unity**, and to employ these few, but important rules, as a series of guide-lines which should help to keep you on track, and moving toward the ultimate goal.

So, in closing, I wish to quote the words of my great friend and mentor, Nikola Tesla:

"Ere many generations pass, our machinery will be driven by a power obtainable at any point in the Universe. This idea is not novel...We find it in the delightful myth of Antheus, who derives power from the earth; we find it among subtle speculations of one of your splendid mathematicians...that throughout space there is energy. Is this Energy static or Kinetic? If static our hopes are in vain; if Kinetic-and we know that it is, for certain-then it is a mere question of time when men will succeed in attaching their machinery to the very wheel work of Nature."

JFMIII 4-9-2018

Dynafux Alternator Patent
Alternator Having Improved Efficiency
US4780632

The Dynaflux Alternator is one of the few low-drag generators to ever be awarded a US Patent. US4780632 was granted on October 25, 1988.

It is important to understand that some details were left out of the patent, but Jim Murray revealed these details in his presentation at the 2015 Energy Science & Technology Conference. The title of the presentation is The Dynaflux Alternator and Lenz's Law, which is available at http://dynafluxalternator.com.

This patent and the latest updated application for the Dynaflux Alternator is included in this book because the Dynaflux Alternator is used in the examples given.

THE MEANING OF UNITY IN ENERGY CONVERSION SYSTEMS

United States Patent [19]
Murray III

[11] Patent Number: 4,780,632
[45] Date of Patent: Oct. 25, 1988

[54] ALTERNATOR HAVING IMPROVED EFFICIENCY

[75] Inventor: James F. Murray III, Piscataway, N.J.

[73] Assignee: MKH Partners, Randolph, N.J.

[21] Appl. No.: 112,025

[22] Filed: Oct. 21, 1987

Related U.S. Application Data

[63] Continuation of Ser. No. 852,995, Apr. 17, 1986.

[51] Int. Cl.⁴ H02K 39/00
[52] U.S. Cl. 310/111; 310/261; 310/216
[58] Field of Search 310/111, 155, 216, 261, 310/266, 171

[56] **References Cited**

U.S. PATENT DOCUMENTS

2,917,699	12/1959	Grant	310/111
3,132,269	5/1964	Craske	310/111
3,321,652	5/1967	Opel	310/168
3,956,649	5/1976	Silverman	310/111
4,639,626	1/1987	McGee	310/168 X
4,659,953	4/1987	Luneau	310/111

FOREIGN PATENT DOCUMENTS

0174290	3/1986	European Pat. Off. .
1538242	10/1969	Fed. Rep. of Germany .
2386181	10/1978	France .
1263176	2/1972	United Kingdom .

Primary Examiner—Mark O. Budd
Assistant Examiner—Judson H. Jones
Attorney, Agent, or Firm—Caesar, Rivise, Bernstein, Cohen & Pokotilow, Ltd.

[57] **ABSTRACT**

An alternator having a rotor which extends obliquely with respect to its axis of rotation between at least a pair of pole pieces having power windings disposed therein with the turns of the windings facing the pivotal axis of the rotor. Upon driving the rotor in rotation, the rotor due to its oblique configuration causes the flux extending with respect to the pole pieces to reciprocate with respect to the pivotal axis and the power windings, thereby producing two cycles of alternating current for each rotation of the rotor. The form of the rotor can substantially correspond to a portion of a cylinder having oppositely disposed face portions extending obliquely with respect to the central axis of the of the cylinder which is substantially coincident with the pivotal axis of the rotor. The periphery of the rotor thereby generates a cylindrical figure of revolution as the rotor is rotated.

32 Claims, 10 Drawing Sheets

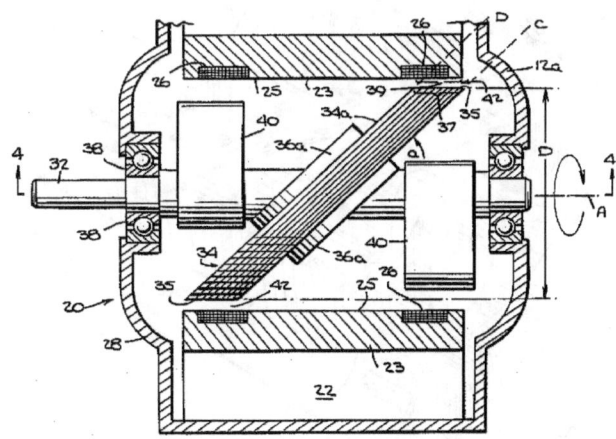

THE MEANING OF UNITY IN ENERGY CONVERSION SYSTEMS

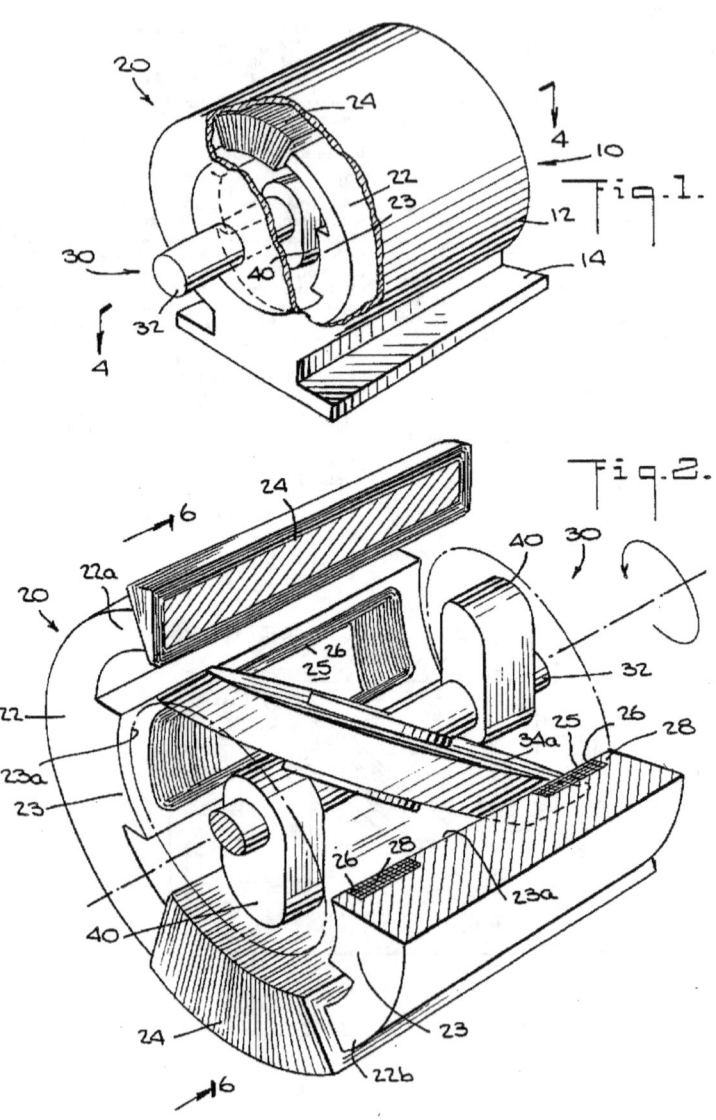

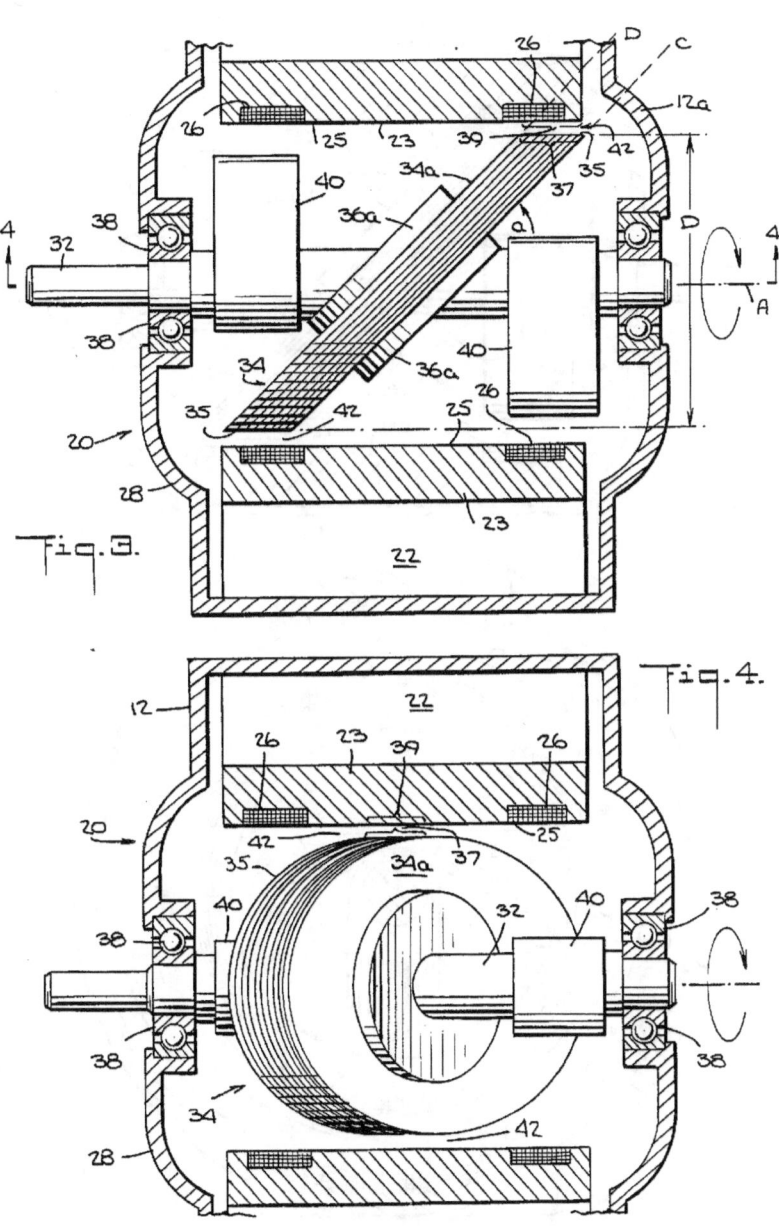

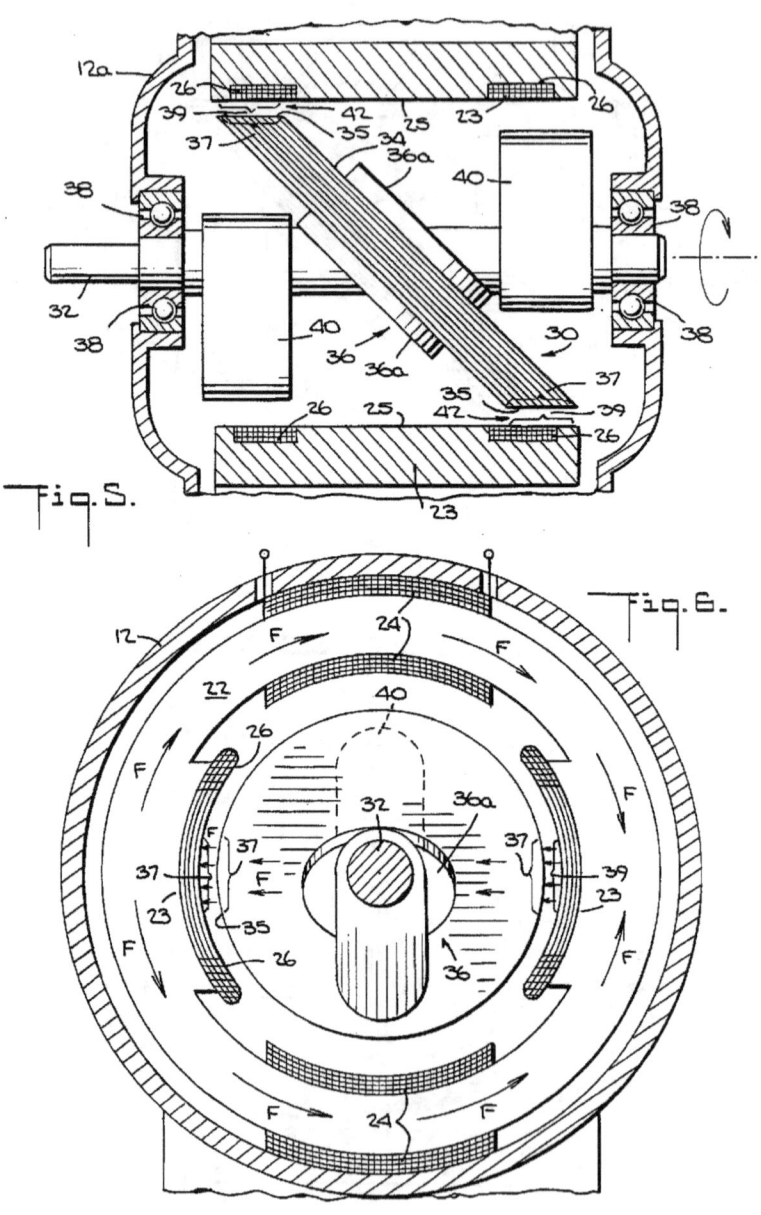

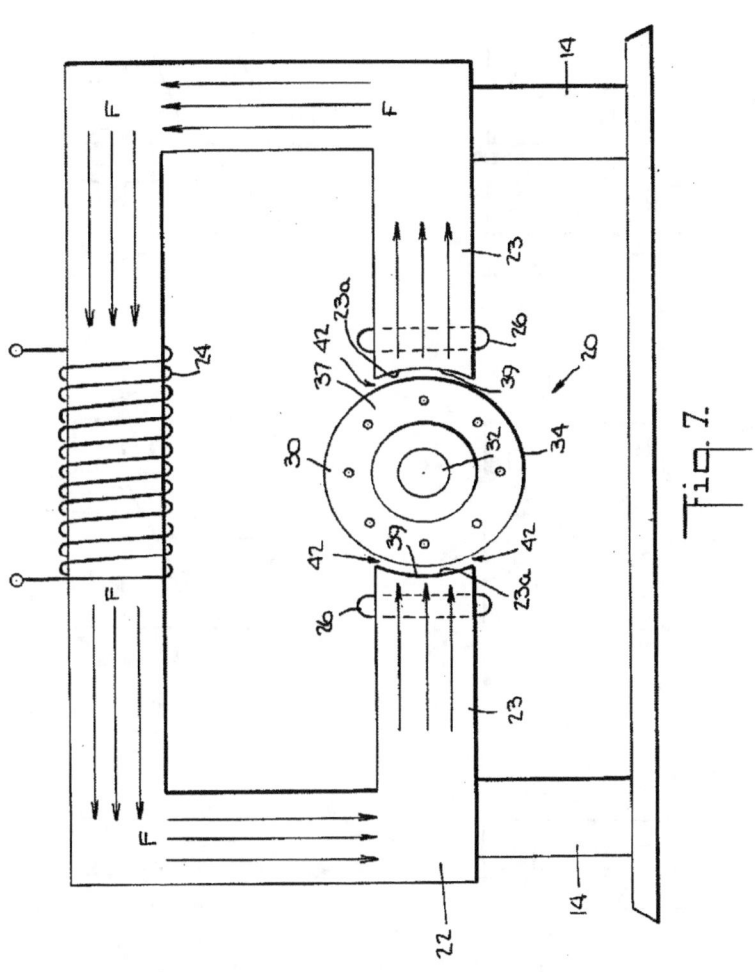

Fig. 7.

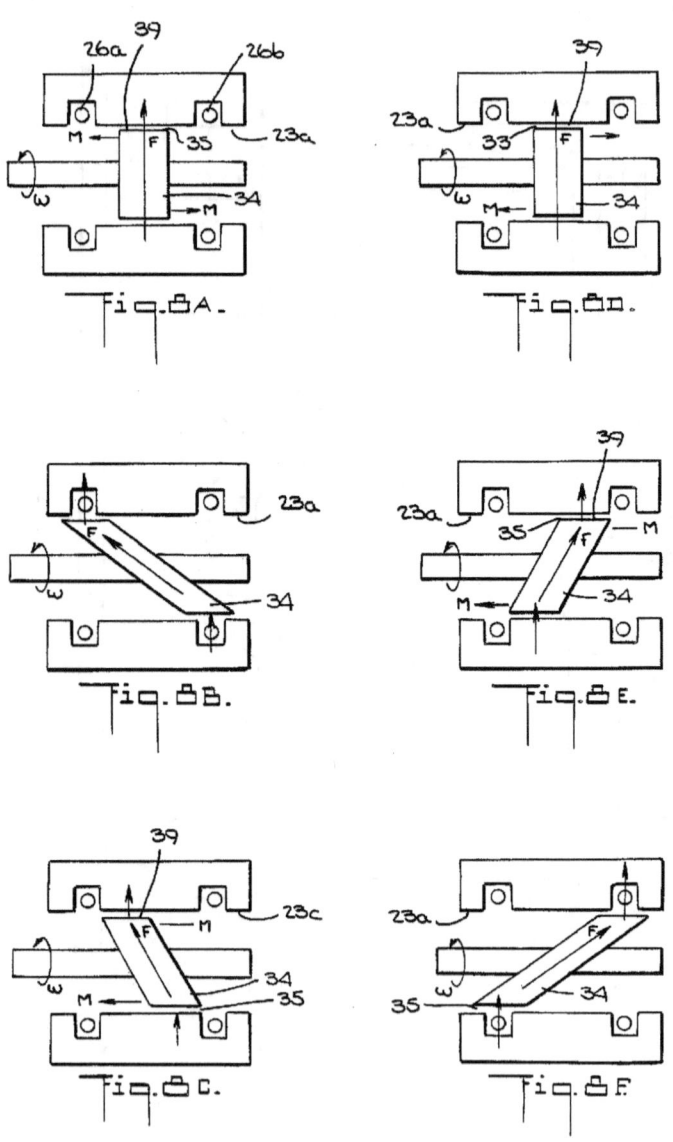

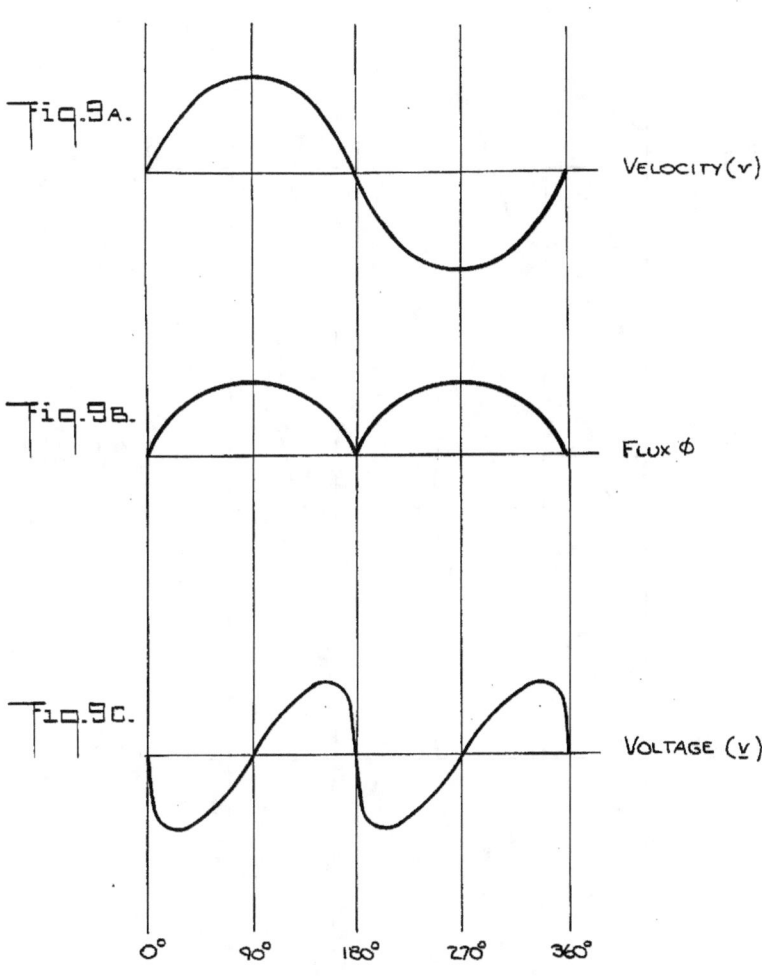

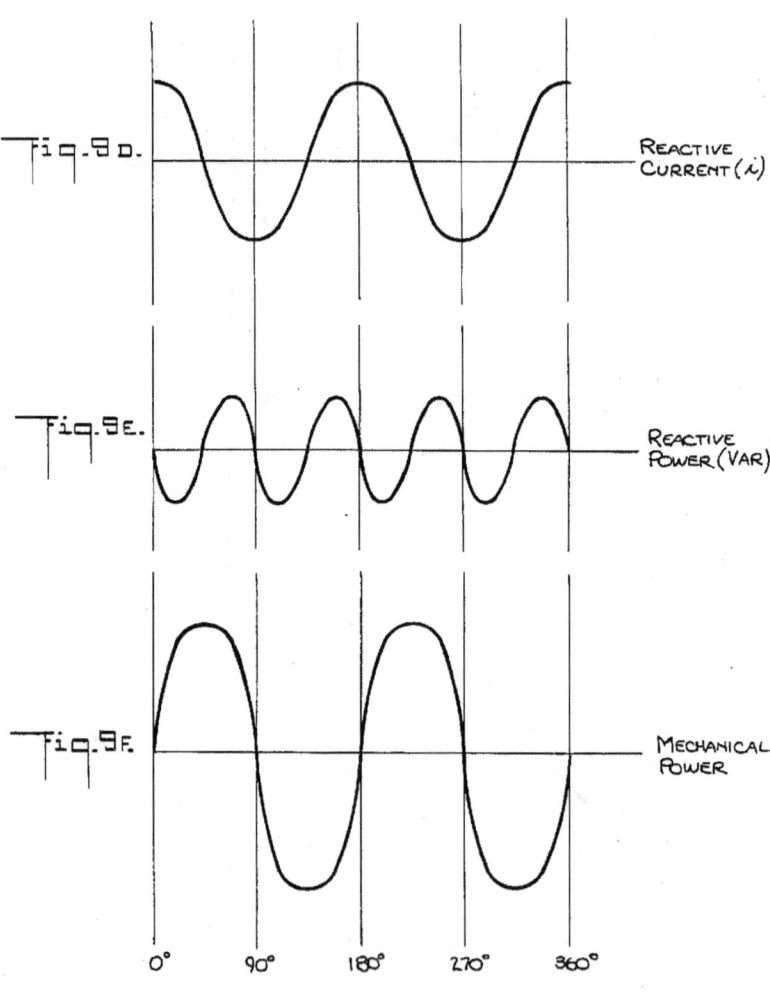

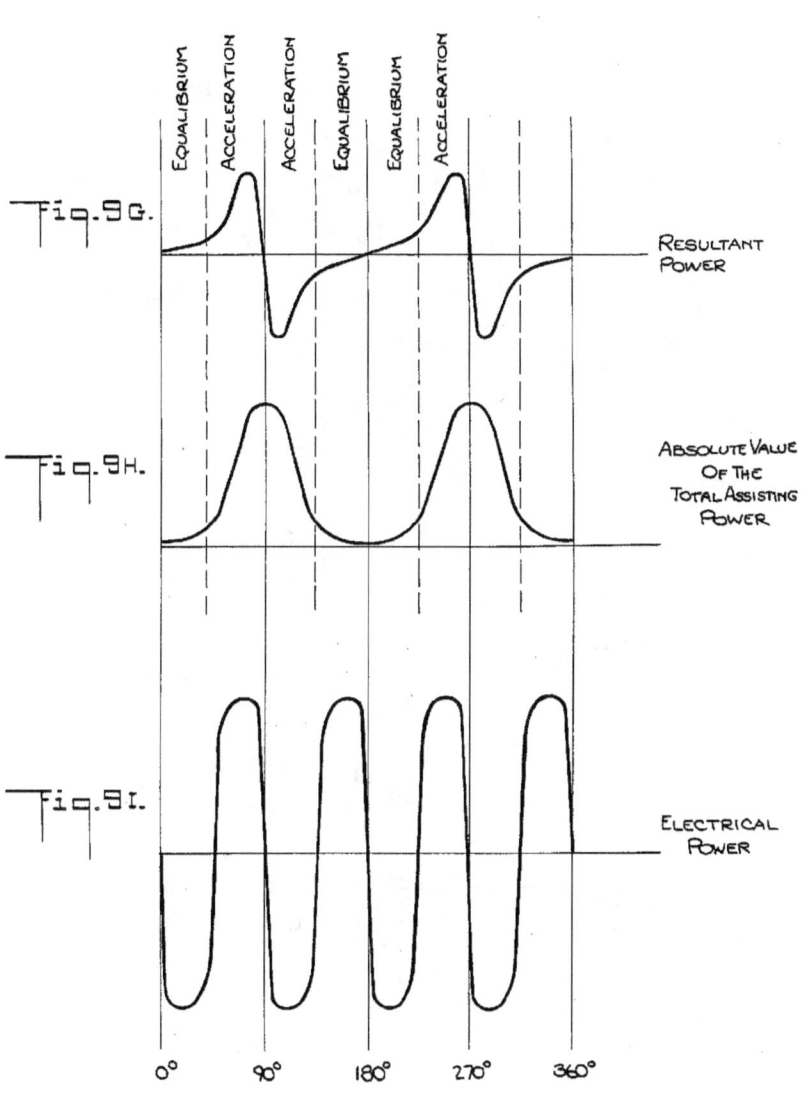

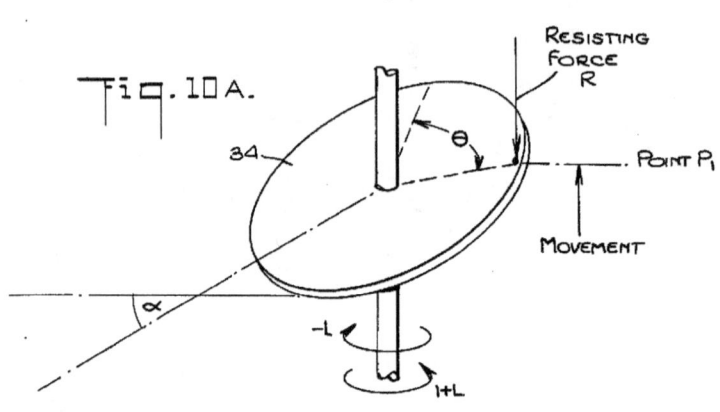

Fig. 10A.

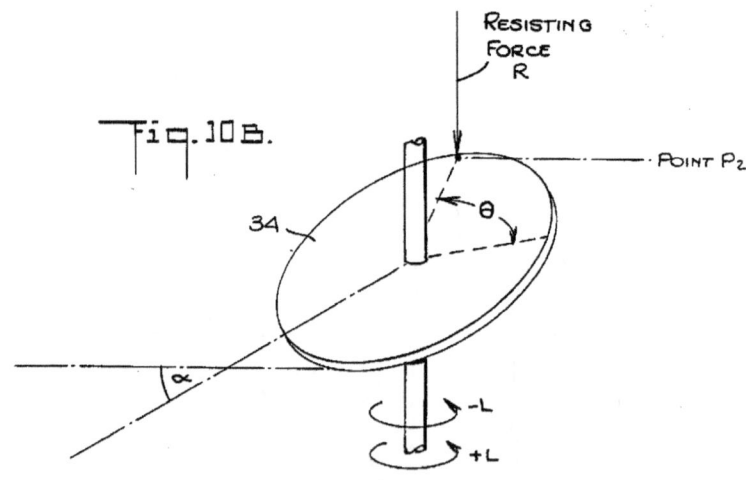

Fig. 10B.

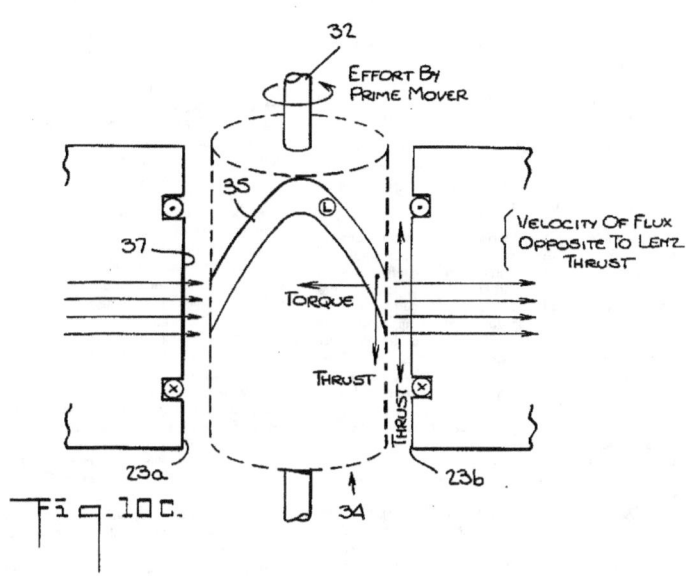

Fig. 10C.

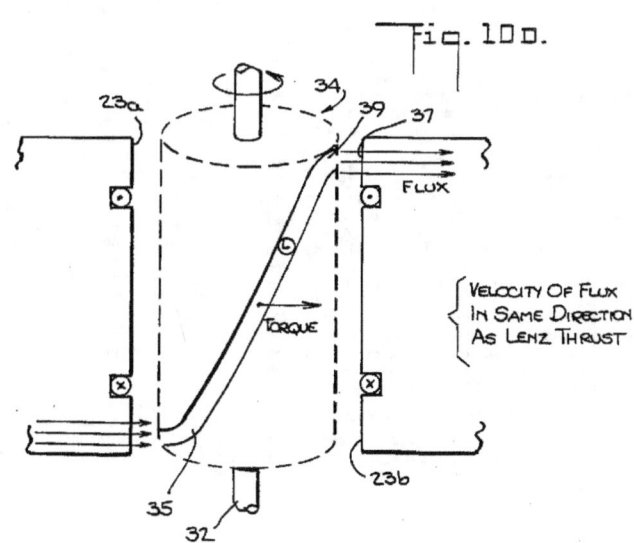

Fig. 10D.

THE MEANING OF UNITY IN ENERGY CONVERSION SYSTEMS

4,780,632

ALTERNATOR HAVING IMPROVED EFFICIENCY

This application is a continuation of application Ser. No. 852,995, filed Apr. 17, 1986.

BACKGROUND OF THE INVENTION

1. Field of the Invention

The invention relates to the field of electrical power generation and more particularly to alternating current generators or alternators. The invention also relates to alternators in which the lateral axis of the rotor is disposed at an oblique angle with respect to the axis of rotation of the rotor. The oblique angle results in the rotor having its angular momentum distributed with respect to two separate axes.

2. Description of the Prior Art

Notwithstanding the increased interest in energy conservation over the last decade, no substantial advance has been made in increasing the efficiency of electrical generating apparatus. Rather, the art has made incremental advances, but in general produces electrical energy with apparatus having approximately the same efficiencies as those used several decades ago.

For example, U.S. Pat. No. 3,321,652, issued to Opel on May 23, 1967, teaches a reduction in windage losses which is achieved by employing a solid rotor having no windings thereon. The rotor incorporates two poles, separated by an area of non-magnetic material. The field coils are coupled to the rotor poles by means of an air gap. The stator windings undergo the same fluctuations in magnetic field as if windings were present on the rotor.

U S. Pat. No. 3,571,639, issued to Tiltins on Mar. 23, 1971, discloses a solid rotor which is made up of alternating magnetic and non-magnetic sections, interleaved by extending finger-like members, and with a magnetic section in the center. This construction provides a two-section alternator capable of increasing the alternator output power for a given shaft speed.

U.S. Pat. No. 3,398,386, which issued to Summerlin on Aug. 20, 1968, also teaches a rotor for a synchro device in which the rotor is without windings and without poles. Rather, the rotor has one face inclined obliquely to the axis of the rotor in order that a point on the stator windings receives magnetic flux in varying strength, depending upon the width of the rotor opposite that point.

Yet another approach is offered by Imris in U.S. Pat. No. 3,760,205 which issued on Sept. 18, 1973. In this patent, the rotor is shaped as a helically-wound flat band which receives magnetic flux through air gaps coupled to the poles of the field magnet. As the rotor rotates, a varying length of the rotor band is coupled to the field source, thus varying the reluctance of the rotor and hence the field strength impressed upon the stator windings.

Therefore, it can be seen that the prior art of alternator constructions despite years of effort has not achieved changes in alternator construction based upon new principles which could dramatically increase the conversion efficiency of the alternator.

SUMMARY OF THE INVENTION

It is therefore an object of the present invention to provide an alternator operating at significantly higher efficiences than those known to the art.

A further object of this invention is to provide an alternator capable of producing a higher-frequency output per shaft revolution per pole set than is known to the art.

Another object of this invention is to provide an alternator which can derive useful work from the reactive power which usually contributes only to the losses in present alternating current generating systems.

These and other objects are achieved in the present invention by providing an alternator with a stator element carrying a combination of field windings and power windings and a rotor which reciprocates the magnetic field across said power windings with substantially simple harmonic motion.

One embodiment of the invention employs a rotor fabricated from a stack of laminated disks, pressed upon an arbor which is obliquely disposed with respect to the intended axis of rotation, and integrally machined in order to provide the assembly with a peripheral contour equivalent to that of a cylinder. The stator is formed of two salient pole projections, each having a concave pole face whose radius is slightly greater than the radius of the rotor. The rotor thereby defines an air gap of continuous dimension when rotated. The rotor is in series with the two pole pieces to complete the magnetic circuit. As the rotor is rotated, the zone at which the flux is coupled to the pole pieces varies in position along the length of each pole face. The magnetic flux is swept forward and then backward across the power windings during each revolution, thereby inducing a voltage therein. Since the magnetic flux reverses its direction of motion, but never its intrinsic polarity, there is a generation of force at the point of reversal which acts in concert with the force provided by the prime mover. Thus, an alternator constructed according to the present invention will produce a given volume of output power for a significantly reduced volume of input power.

BRIEF DESCRIPTION OF THE DRAWINGS

FIG. 1 is a perspective view of a preferred embodiment of the alternator of the invention, partially cut away to illustrate interior components thereof;

FIG. 2 is a perspective view of the stator and rotor assemblies of the embodiment shown in FIG. 1, cut away to show interior components thereof;

FIG. 3 is a horizontal section of the embodiment of the invention shown in FIG. 1 with the maximum dimension of the rotor disposed adjacent to the end portions of the pole pieces;

FIG. 4 is a horizontal section of the embodiment of the invention shown in FIG. 1 with the minimum dimension of the rotor disposed adjacent to the central portions of the pole pieces;

FIG. 5 is a horizontal section of the embodiment of the invention shown in FIG. 3 with the rotor turned 180 in the direction of the arrow from the position shown in FIG. 3;

FIG. 6 is a vertical section taken along the line 6—6 in FIG. 2;

FIG. 7 is a schematic representation of the magnetic circuit of the alternator of the invention.

FIGS. 8A-8F are schematic representations of horizontal sections of the invention showing the magnetic flux between the stator and rotor for six different rotational positions for one rotation of the rotor;

64

FIGS. 9A-9I are graphical representations of nine electrical and mechanical characteristics for a complete rotation of the rotor; and

FIGS. 10A-10D are schematic representations depicting the interaction of magnetic and mechanical forces within the alternator of the invention.

DETAILED DESCRIPTION OF A PREFERRED EMBODIMENT

FIGS. 1 and 7 show a preferred embodiment of the alternator 20 of the invention. The alternator comprises a stator assembly 22 and a rotor assembly 30 disposed within a housing 12. The housing is supported by a base member 14.

In FIG. 2 it can be seen that the stator assembly 22 has the general form of a hollow cylinder. The stator assembly is formed of a highly permeable material and is provided with two pole pieces 23 which extend radially inwardly and terminate in concave faces 23a.

The stator assembly 22 carries two sets of windings. Field windings 24 may be carried on the stator body in a convenient location, for example, at the top portion 22a and the bottom portion 22b of the stator iron as shown in FIG. 2. The constructional details of such windings are well-known to the art.

Power windings 26 are also carried by the stator assembly with one or more windings on each pole piece 23. (FIGS. 2 and 7). The windings 26 are located in slots 28 extending in to the face 23a of each pole piece 23. The slots should be of sufficient depth to insure that the windings 26 disposed in them do not protrude into the air gap 42 (FIGS. 3 and 7). It should be noted that the embodiment of the invention, i.e., alternator 20, comprises one pair of pole pieces 23; however the alternator of the invention can be constructed in embodiments containing multiple pairs of poles.

The rotor assembly 30 can best be seen in FIGS. 2, 3, and 6. A shaft 32 carries a rotor 34 within the hollow cylinder defined by the stator body. The shaft is journaled on suitable bearings 38 (FIG. 3) mounted in each of the opposite ends 12a of the casing 28. A prime mover (not shown), is connected to shaft 32 to provide a driving torque. The rotor 34 is fabricated from material having high permeability, preferably, electrical steel, such as a stack of silicon steel laminations to reduce or minimize eddy currents. The rotor 34 is secured to shaft 32 by an arbor 36a. Counterweights 40 can be mounted on the shaft 32 to provide a balanced mechanical structure.

The form of the rotor 34 as shown in FIG. 3, is a section of a cylinder having a diameter D and an axis A, which is cut by two parallel planes "B" and "C". In the preferred embodiment, angle "a" is 45°. FIG. 3 shows the rotor at the beginning of a cycle of rotation when the rotor is seen as if on edge. FIG. 4 shows the rotor viewed 90° from the position of FIG. 3. In this position, the face 34a of the rotor can be seen to be elliptical. FIG. 5 shows the rotor after a movement of 180° from the position shown in FIG. 3.

The areas on the rotor edge 35 and a portion of the faces of the pole pieces 23 are referred to as coupling zones 37 and pole face flux zones 39, respectively (FIGS. 3-6, 10C and 10D). The pole face flux zones oscillate along the length of each pole face with simple harmonic motion as the rotor assembly is revolved. Thus, the position of the upper pole flux zone 39 as shown in FIG. 3 is located at the right-hand end of the pole face, while the same zone as shown in FIG. 4 is located at the center of symmetry of the power winding. In FIG. 5, this zone has travelled to the left-hand end of the pole face. Thus, as the rotor stack turns through the next 180°, the pole face flux zones 39 return to the position shown FIG. 3. These zones execute harmonic motion with respect to

Operation of the alternator according to the present invention requires the presence of a d.c. excitation current in the field windings 24 and the application of a torque to shaft 32. The current flowing in the field windings 26 produces a stationary magnetic field in the stator iron 22 with the lines of flux tending to flow in the magnetic circuit by following the path of least reluctance, as illustrated by arrows F in FIG. 6 and as shown in FIG. 7. Flux will flow through the stator 22 to the flux zone 39 of pole pieces 23. From there, flux will pass across the air gap 42 to the flux zone 37 of the rotor 34, returning across the air gap 42 to the pole piece of opposite polarity, and then back to the stator. Thus, the rotor magnetically couples the two pole faces 23a, by providing a low-reluctance path between the pole pieces. Since the peripheral portions of the rotor are parallel to the pole faces, a maximum flow of flux will be obtained across the entire thickness of the rotor.

A more detailed understanding of the operation of this invention can be gained from FIGS. 8A-8F. FIGS. 8A-8F are each a schematic representation of a cross-section of a face 23a of a pole piece face and the opposing rotor edge portion 35 of the rotor 34, taken at different points during one cycle of rotation of the rotor. FIG. 8A depicts a point in the rotor cycle at which the flux zone 39 is located midway between the conductors of power winding segments 26a and 26b. As noted above, the rotor geometry results in the flux zones being moved reciprocally back and forth across the faces of the pole pieces. As shown in FIG. 8A, the rotor edge (and thus, the flux zone 39) is being accelerated in the direction of arrow M, toward power winding segment 26a.

FIG. 8B depicts the situation after 90° of rotor rotation in the direction of the arrow. Here, the flux zone has moved to overlap power winding segment 26a with rate of change of flux becoming zero. Accordingly, the voltage becomes practically zero. This is the location at which the point of reversal in direction of the flux zone takes place. FIG. 8C shows the situation after another 45° of rotation. The pole face flux zone has returned part way toward the midpoint of the power winding in a direction extending toward power winding segment 26b.

FIG. 8D depicts the condition at 90° of rotation with the pole face flux zone 39 at the midpoint between winding segments 26a and 26b.

In FIG. 8E there is shown the condition after another 45° of rotation.

In FIG. 8F, the pole flux zone is at an extreme point of movement relative to the pole piece face and including power winding segment 26(b). No directional arrow is shown since the zone is momentarily at rest with respect to the pole piece face. In this way, a full cycle is completed.

FIGS. 9A-9I graphically shows the operation of the alternator of the invention in terms of the relevant mechanical and electrical parameters. Each graphic function shown in FIGS. 9A-9I relate by degree markings to points in a single cycle of rotation of the rotor. The velocity of the pole face flux zones with respect to a point on the pole face is plotted at FIG. 9A. As a result of geometry of the rotor, the velocity is sinusoidal in form.

Flux (ϕ) is plotted in FIG. 9B. As shown, the relative flux concentration alternates from zero to a maximum value twice each cycle without undergoing a reversal of its magnetic polarity. The flux does not vary sinusoidally, but exhibits a complex harmonic variation which can be expressed mathematically as a Fourier series as follows:

$$\phi = \phi m(1 + 2/3 \cos 2\ wt - 2/15 \cos 4\ wt + \tfrac{8}{?} \cos 6\ wt + \ldots)$$

where ϕm is the maximum flux.

The expression above reveals an important advantage of the present invention. As shown by the appearance of the number "1" in the flux equation, the flux carries a component analagous to a "direct current component" of a complex electrical waveform, caused by the fact that the flux never reverses polarity. Under these conditions, the iron domains within the pole pieces will not exhibit major hysteresis loops usually associated with oscillating flux. Thus, the present invention drastically reduces hysteresis losses which are particular to all flux-reversing systems. In addition, cooling requirements of the alternator are likewise reduced since smaller quantites of heat are generated by the reduced hysteresis losses in the pole pieces.

The voltage induced in the power windings is proportional to the rate of change of flux and may be calculated by the equation $V = -d\phi/dt$. Since the flux does not vary sinusoidally, the voltage will have a non-sinusoidal waveform. Differentiation of the Fourier series shown above results in an expression for the instantaneous voltage:

$$V = \phi m\ (4/3 \sin 2\ wt - 8/15 \sin 4\ wt + 12/35 \sin 6\ wt + \ldots).$$

This function is plotted in FIG. 9C. Note that the induced voltage oscillates through a complete cycle for every 180° of rotation of the rotor. Thus, the induced voltage has twice the frequency of the harmonic velocity of the flux and the angular velocity of the shaft. This fact has an important consequence, as the prior art teaches that a two-pole alternator can generate only one cycle of current for each revolution of the rotor. Thus, the prior art requires that a two-pole alternator must operate at 3600 r.p.m. in order to generate 60 cycle alternating current. By means of the alternator of the invention, a two-pole machine can produce 60 cycle alternating current at 1800 r.p.m.

The alternator of the invention also results in a reduction of iron-related losses which are proportional to rotor speed as well as loses stemming from mechanical friction and windage. Additionally, a reduction in shaft speed can offer increased reliability and longer life due to the reduction of wear in mechanical parts.

In an inductive circuit, such as that of the power windings of an alternator, it is well-known that a certain component of the current flows in a reactive relationship to the induced voltage. This "produces" a "reactive power component" referred to as "volt-amperes reactive", or "VAR" power. The average value of reactive power is zero, and it can make no contribution to the consumed power such as a resistive load. However, due to the fact that the flux changes its direction mechanically, the energy stored stored in the VAR component can be transformed into useful mechanical work, and assist the prime mover in rotating the rotor shaft. The prime mover must perform work to turn the rotor shaft and generate electricity. This phenomenon is the specific consequence of the more general Lenz' Law to an alternator.

The maximum power transfer theorem states that: "maximum power transfer between source and load occurs when the load impedance is the conjugate of the source impedance, and under these conditions, the source can transfer only 50 percent of the converted power to the load. In an alternator, the true EMF resides as a spatial vector referred to as a "motional" electric field. The total power dissipated in the source windings can be measured directly only in terms of the mechanical torque supplied to the generator's drive shaft as would be expected according to the dictates of the Lenz Law. The alternator's electrical phase angle, which is a "space" angle, is a measurement of the angular position of the conductors in space relative to the position of the generator's magnetic field poles when compared to the induced voltage. The oscillating VAR component is transformed into a mechanical tortional vibration residing in the generator's armature mass and drive shaft. This fact indicates that the alternator attempts to return the reactive power to its own source, the prime mover, on alternate quarter cycles. In an attempt to maximize the delivery of power to the load in accordance with the maximum power transfer theorem, the load impedance can be made conjugate to the power impedance by the addition of capacitance.

The maximum power transfer theorem describes the condition necessary to ensure optimal transmission of power from source to load. This theorem can be applied to power transmission theory, however, its main utilization is in the field of radio and telephone communications. The alternator of the invention utilizes a reverse power transfer concept, which was developed solely for alternating power applications, particularly those of a single phase nature, which involve specially designed elliptical rotor geometries.

The basis of this concept can best be grasped by referring to FIG. 10C. This drawing shows an oval rotor 34 pictured within its cylindrical surface of revolution. At the instant depicted, the rotor 34 is so positioned that the flux is centered on each pole face, and is passing through the axis of symmetry of the lamination stack. As rotation proceeds, from left to right, the flux in the left pole face is moved in a downward direction, and begins to induce a voltage in 23a, the flux in the right pole face is moved in an upward direction and begins to induce a voltage in 23b. Assume for simplicity and by way of example, that the coils are connected in additive series, and that their output is short circuited. This will ensure that the windings are the only active components in the circuit, and that the power produced in them will be substantially reactive.

As current starts to build within the coils, an opposing force due to the Lenz reaction will attempt to thrust the flux in a direction opposite to that of its motion. This thrust will be parallel to the axis of rotation of shaft 32, and in an opposite sense for each pole. The action of these forces upon the lamination stack will be analogous to that of followers in the groove of a cylindrical cam. Hence, these lateral thrusts will be converted into torques which oppose the effort of the prime mover for one quarter cycle.

THE MEANING OF UNITY IN ENERGY CONVERSION SYSTEMS

In FIG. 10A, a resisting force is shown by an arrow and identified by the letter R as being applied to point P_1 on the rotor. Another arrow marked $-L$ represents the torque created by this resisting force R. The arrow marked $+L$ represents the effort exerted by the prime mover. The system is in dynamic equilibrium as shown in FIG. 10A where torque $-L$ is equal and opposite to torque $+L$.

In FIG. 10B, the arrow marked $-L$ again represents the torque created by the resisting force R now applied to point P_2. The arrow marked $+L$ again represents the effort exerted by the prime mover. Since torque $+L$ and torque $-L$ are in the same direction, the system now accelerates.

If the resisting force R is moved from P_1 to P_2, the sense of $-L$ is reversed. Note that the disk has not been rotated, but the point of contact has been moved through an angle ϕ. If the disk had been rotated and the resisting force remained fixed in space, the relative results would be the same, i.e., an assisting torque would suddenly appear after turning through the angle θ.

In FIG. 10D, the drive shaft 32 has rotated 90 mechanical degrees, and the lamination stack has traversed a space angle of 90 degrees relative to the pole faces. At the instant depicted, the harmonic velocity of the surface of the lamination stack relative to the pole faces is exactly zero, but about to reverse. At this point in time, the reactive current in each winding is just reaching its maximum value because it is 90 degrees out of phase with the induced voltage.

Hence, as each edge of the lamination stack begins to accelerate in the opposite direction, relative to the pole faces, magnetic forces produced by the current in each winding now attract the flux, and develop thrusts which operate in the same direction as that of the motion. Due to the cam-like design of the lamination stack, these actions give rise to torques which now assist the effort of the prime mover for the next quarter cycle.

This exchange of energy between the magnetic field of the inductor and the mass of the rotor 34 constitutes a form of resonance, which hereinafter is referred to as energy resonance, and which is the underlying principle in the concept of reverse power transfer.

Referring now to FIGS. 9A-9I, the relationships between the various mechanical power constituents and the magnetic flux can be studied. It should be noted that the velocity (v) shown in FIG. 9A is a sine function, and is in phase with the flux shown in FIG. 9B (ϕ) for $\frac{1}{2}$ a cycle and then 180 degrees out of phase with the flux for the next $\frac{1}{2}$ cycle. Since the force (F) is a cosine function, the product of the velocity and the force produce a sinusodial wave of mechanical power (Pm) shown in FIG. 9F which oscillates at twice the frequency of the velocity. It must be understood that this wave function is representative of the mechanical power relative to the magnetic pole faces only. The power on the shaft as seen by the prime mover is a far more complex function.

The constituents of the electrical power can now be examined. Notice that the voltage (V) shown in FIG. 9C is twice the frequency of the velocity. Because of the inductive nature of the circuit of the alternator being operated into a short circuit in the example referred to above when coils are connected in additive series, the current (I) lags the voltage by 90 degrees. The product of the voltage and the current yields a negative sine function with twice the frequency of the mechanical power. It must be understood that the (R_R) curve of FIG. 9I is representative of the reverse power transfer function, and as such, it can be interpreted as electrical power or mechanical power depending upon which side of the conversion boundry the observer chooses to focus his attention.

It is here assumed that the curve of (P_R) FIG. 9I represents the reverse power function on the mechanical side of the conversion boundry. This places both power waves in the mechanical domain and it immediately becomes clear that a "beat" will be established between the two frequencies. (FIGS. 9F and 9I).

Referring to FIGS. 9A-9I, it will be seen that for the first quarter cycle, the mechanical power (Pm) is 180 degrees out of phase with the reverse power (R_R). This is indicative of power flowing from the prime mover through the conversion barrier and into the magnetic fields of the power coils. The average values of these two wave fragments cancel, and result in nearly zero power in the first quadrant of the resultant power curve seen in FIG. 9G when the system is in equilibrium.

Just after 90 degrees, the reverse power function in FIG. 9I starts to go positive, and is now in phase with the second 90 degree portion of the mechanical power curve. This is indicative of power being returned to the prime mover from the magnetic fields of the power coils; however, this action coincides with the positive acceleration cycle of the lamination stack. Hence, the average values of these wave fragments reinforce, resulting in the high amplitude of the second quadrant of the resultant power curve seen in FIG. 9G. This curve is a second harmonic periodic function, and as such can cycle repeatedly.

If it were not for various losses which absorb most of the reflected power, this type of hetrodyne wave amplification could cause the system to accelerate to destruction if operated in the short circuit mode, as suggested in this example, i.e., pure mechanical resonance; however, with losses being numerous and rather high in electro-mechanical equipment, the net result is not acceleration, but rather a diminution of the drive torque demanded by the alternator from the prime mover. This is a typical example of applied energy resonance. In actual practice, alternators are not short circuited, but are used to deliver power to useful loads.

Applying a resistive load to the output of the alternator of the invention causes a phase shift to occur between the mechanical power wave, and the reverse power wave, such that only a portion of the stored inductive energy is fed back from the alternator's power coils to the rotor's momentum; however, if the losses in the alternator have been accurately anticipated, they can be greatly compensated for by choosing the proper ratio of reactance to the intended kilowatt load. This implies that the efficiency of the alternator of the invention is directly proportional to the Q ratio in this energy resonant system.

Should the inductance of the power coils be frequency resonated as suggested by the maximum energy transfer theorem by the introduction of the proper capacitance to the external circuit, the current will immediately fall into phase with the induced voltage, with the overall result that the reverse power transfer wave will degenerate into a sine squared function in phase with the mechanical power. This will destroy the bidirectional power coupling of the system by critically dampening the energy resonance. The result would then be that the alternator of the invention would behave similarly to a conventional alternator.

THE MEANING OF UNITY IN ENERGY CONVERSION SYSTEMS

In summary, immediately upon a reversal after the condition shown in FIG. 8B, the flux zone has begun movement to the right toward the condition shown in FIG. 8C. The flux is increasing and therefore the induced voltage has changed direction. The inductive current component, however, continues to flow in the same direction that it did in FIGS. 8A and 8B. Thus, the force associated with the inductive current continues to be exerted and this thrust is now in cooperation, not in opposition, to the movement of the rotor. Therefore, this force no longer produces a torque in opposition to that exerted by the prime mover, but one that assists the prime mover in turning the rotor. Since this phenomenon occurs for a portion of each cycle, immediately following a reversal in direction of the flux zone'e harmonic motion, a prime mover when driving the alternator of the invention will have to exert considerably less effort to turn the rotor shaft than does a prime mover driving a conventional alternator.

As a result, the alternator of the invention exhibits a dramatically increased efficiency of operation, there being a larger ratio of power-out to power-in. The net effect of this phenomenon can be estimated theoretically by taking the ratio of the electrical power available to the mechanical power normally needed to turn the rotor shaft.

A demonstration of the existence and efficacy of the phenomenon described herein emerges from an examination of Tables 1 and 2. Table 1 shows test data for a standard, commercially available alternator while Table 2 shows test data for an alternator produced according to the teachings of the present invention. Each alternator produces power at 60 hertz. Each device was tested using the IEEE Standard Test Procedure for Synchronous Machines, S #115-1983. Those familiar with the art will recognize the definitions and procedures.

TABLE 1
Standard Prior Art Alternator
Segregated Losses & Power Accounting
(in watts at 60 cycles)

Field Voltage A	Total Input Power Watts B	Friction & Windage Losses at 3600 RPM C	Actual Input Power D	Iron Losses E	Stray Copper Losses F	Resistive Load Wattage G	Total Converted Power H (E + F + G)	Percent Account of Input Energy I (H/D × 100)
0	232	232	0.	0.	0.	0.	0.	—
1	236	232	4.00	0.500	2.50	.536	3.536	88.40%
2	240	232	11.00	2.00	3.75	2.162	7.912	71.92%
3	246	232	14.00	4.00	4.90	4.260	13.160	94.00%
4	252.50	232	24.51	6.00	6.35	7.395	19.745	80.55%
5	261.33	232	29.33	9.00	7.50	10.640	27.140	92.53%
6	268.21	232	36.21	12.50	8.75	14.340	35.590	92.28%
7	278.77	232	46.77	16.00	10.00	18.870	44.870	95.93%
8	288.00	232	56.00	20.00	11.25	23.125	54.375	97.09%
9	298.65	232	66.65	24.00	12.50	27.533	64.033	96.07%
10	309.40	232	77.40	28.00	13.75	33.010	74.760	96.58%
11	318.75	232	86.75	32.00	14.90	37.350	84.250	97.11%
12	327.25	232	95.25	36.00	16.25	41.460	93.710	98.38%
13	338.40	232	106.40	40.00	17.50	46.140	103.640	97.40%
14	346.75	232	114.75	43.00	18.75	50.410	112.160	97.74%

TABLE 2
Alternator of Invention
Segregated Losses & Power Accounting
(in Watts at 60 Cycles)

Field Voltage A	Total Input Power B	Friction & Windage Losses at 1800 RPM C	Actual Input Power D	Iron Losses E	Stray Copper Losses F	Resistive Load Wattage G	Total Converted Power H	Apparent Excess Power J (D − H)	VAR Power K	Total Input Register L (H − K)	Percent Account I
14	130.50	109.58	20.92	20.42	4.80	2.471	27.691	−6.771	7.102	20.588	98.41%
16	137.96	109.58	28.38	25.42	6.66	4.070	36.150	−7.770	9.919	26.231	92.42%
20	154.78	109.58	45.20	40.34	10.00	6.826	57.166	−11.966	14.652	42.514	94.05%
24	171.00	109.58	61.42	57.92	12.50	9.758	80.178	−18.758	19.229	60.949	99.23%
26	182.72	109.58	73.14	67.22	14.80	11.359	93.379	−20.239	21.750	71.629	97.93%
28	195.50	109.58	85.92	76.50	15.20	13.289	104.989	−19.069	24.188	80.801	94.04%
30	208.86	109.58	99.28	86.34	17.20	15.411	118.951	−19.671	26.629	92.322	92.99%
32	220.65	109.58	111.07	96.58	18.50	16.945	132.025	−20.955	28.408	103.617	93.28%
34	232.00	109.58	122.42	107.92	20.50	18.852	147.272	−24.852	31.050	116.222	94.93%
36	245.66	109.58	136.08	118.47	21.33	20.390	160.190	−24.110	32.414	127.776	93.89%
38	259.30	109.58	149.72	129.17	22.50	21.896	173.566	−23.846	33.800	139.766	93.35%
40	271.55	109.58	161.97	140.42	24.00	23.583	187.553	−25.583	35.954	131.599	93.59%
42	283.23	109.58	173.65	152.12	25.20	25.080	202.40	−28.750	37.570	164.83	94.92%
44	294.17	109.58	184.59	161.82	26.66	26.419	214.899	−30.309	39.070	175.829	95.25%
46	307.28	109.58	197.70	171.64	27.50	27.887	227.027	−29.327	40.599	186.428	94.29%
48	318.76	109.58	209.18	181.79	29.33	29.583	240.703	−31.523	42.789	197.914	94.61%
50	328.38	109.58	218.80	190.26	31.00	30.992	252.252	−33.452	44.388	207.864	95.00%
52	338.15	109.58	228.57	197.92	32.00	31.979	261.899	−33.329	45.362	216.537	94.73%
54	346.16	109.58	236.58	204.22	32.50	33.040	261.760	−33.180	46.017	223.743	94.57%
56	356.26	109.58	246.68	215.43	33.33	34.226	282.986	−36.306	47.009	235.977	95.66%
58	365.40	109.58	255.82	219.82	35.00	35.976	290.796	−34.976	48.684	242.112	94.64%

THE MEANING OF UNITY IN ENERGY CONVERSION SYSTEMS

4,780,632

TABLE 2-continued

Alternator of Invention
Segregated Losses & Power Accounting
(in Watts at 60 Cycles)

Field Voltage A	Total Input Power B	Friction & Windage Losses at 1800 RPM C	Actual Input Power D	Iron Losses E	Stray Copper Losses F	Resistive Load Wattage G	Total Converted Power H	Apparent Excess Power J (D − H)	VAR Power K	Total Input Register L (H − K)	Percent Account I
60	374.65	109.58	265.07	228.92	35.50	36.433	300.853	−35.783	49.363	251.490	94.87%

The tests were conducted by driving each alternator at its operating speed (3600 r.p.m. and 1800 r.p.m., respectively), applying a given d.c. voltage to the field winding, and employing several standard loading techniques. Total input power was measured, and losses were segregated into friction and windage losses, iron losses, and stray copper losses, according to the cited Standard. Finally, the resistive load wattage was measured. With this data, "Actual Input Power" can be calculated after subtracting the purely mechanical losses of friction and windage. It follows directly that this "Actual Input Power" must appear either as a loss (iron losses or stray copper losses) or as power delivered to the resistive load. The accuracy of the procedure can be determined by comparing the total of all losses and delivered power with the "Actual Input Power."

In Table 1 and Table 2, the values of the columns are defined below with all power values in watts:
A—voltage applied to field windings
B—total input power of prime mover
C—friction and windage loss at 3600 RPM
D—actual input power to alternator (B-C)
E—iron losses
F—stray copper losses
G—resistive load wattage
H—total converted power (E+F+G)
I—percent account of input energy (H/D×100)
J—apparent excess power-(D-H)
K—VAR power
L—total input register-(H-K)

An inspection of Table 1, the data obtained from testing a conventional prior art alternator, reveals that this test procedure thoroughly explains the workings of a standard device. Although power determinations are difficult at low levels of field voltage, when one arrives at normal working levels of field voltage, one can consistently account for the allocation or consumption of more than 95% of input power in terms of the stated categories. This is shown by column I, "Percent Account of Input Energy".

Table 2 which is based upon operating data relating to the invention, validates the theoretical presentation above by demonstrating that an alternator according to the invention performs in a radically different manner from a conventional prior art alternator. Specifically, the explanation for the data contained in Table 2 is that the reactive power stored in the power windings is contributing to the power required to turn the rotor. The data for Table 2 were gathered in the same manner as that for Table 1, using the cited Standard. The first point that can be made is that 60-cycle power was produced from this alternator at 1800 r.p.m., rather than the 3600 r.p.m. that conventional teaching would mandate.

As with the standard alternator, actual input power (column D) was measured, and iron losses (column E), stray copper losses (column F), and resistive load wattage (column G) for the alternator of the invention were measured as shown in Table 2. It was found that when one added the loss categories to the resistive load wattage (column G) to obtain a "Total Converted Power" (column H), that total exceeded the "Actual Input Power" (column D). Thus for every level of field voltage tested, the amount obtained for "Total Converted Power" (column H) exceeds the "Actual Input Power" (column D), as reflected in column J labeled "Apparent Excess Power." Of course, this "excess" power must be derived from some source, and the answer is that some of the reactive power stored in the alternator's power winding is being applied to the rotor, as explained above, thereby reducing the work performed by the prime mover. The amount of power so derived is reflected in column K labeled "VAR Power".

For purposes of determining accuracy of the test measurements, the "Actual Input Power" (column D) is added to the absolute value of the "VAR Power" (column K) and the "Apparent Excess Power" (column J) to obtain a "Total Input Register" (column L), which is then compared with the "Total Converted Power" value (column H) to determine the amount of energy actually accounted for. It can be seen that the percent account of input energy (column I) is well over 90%, which is testimony of the accuracy of the measurements shown in Table 2.

The alternator of the invention can be operated as a synchronous alternating current motor. When the power windings, i.e. stator windings are connected to an alternating current source, there is a reaction between the armature currents and the air-gap flux which results in a torque when the alternating current source is single phase. The torque tends to drive the rotor one way and then the other. Accordingly the rotor will not run; however if the rotor is driven externally into synchronism with any source of mechanical power, a torque will then be produced that is constant in value and is continually in one direction. This torque will keep the motor running at synchronous speed, even under load.

In the case of a single phase source of alternating current connected to a single phase motor embodiment of the invention, when the stator windings are energized and the rotor is stationary, a constantly reversing torque is produced. Thus, the rotor will not start; however, if the rotor is externally driven to synchronous speed, current flowing in the rotor will then react with the field flux to produce an average torque which is always in one direction. Since the current in the rotor changes as the rotor turns, the instantaneous torque is not constant as in a polyphase motor but pulsates.

It should be understood that the embodiment discussed above and depicted in the drawings is for illustrative purposes only, and that those having skill in the art will understand that modifications and alternations can be made thereto within the spirit of the present

What is claimed is:

1. An electrical generation device comprising:
a rotor assembly having an axially extending, rotatable shaft, a rotor affixed obliquely to the shaft and including an outer peripheral edge having an axial dimension provided between axially spaced-apart end faces of said rotor, said end faces being obliquely oriented relative to the shaft, circumferentially spaced segments of said outer peripheral edge being in different axial positions along the shaft as a result of the oblique orientation of the rotor to the shaft;
an external magnetic field circuit having at least first and second spaced-apart pole pieces, means for providing a magnetic field extending through the external magnetic field circuit for establishing a single magnetic polarity in each pole piece, with the magnetic poles of said at least first and second spaced-apart pole pieces being of opposite polarities, each of the pole pieces being of an axial dimension greater than the axial dimension of the outer peripheral edge of the rotor and having a face with a plurality of slots therein, and a plurality of power windings disposed in the slots; and
means for mounting the rotor assembly for relative motion with respect to the first and second pole pieces of the external magnetic field circuit by rotatably mounting the shaft relative to the external magnetic field circuit such that circumferentially spaced segments of the pheripheral edge of the rotor, in axial section, are substantially parallel to and spaced from confronting faces of said first and second spaced-apart pole pieces for defining an air gap between said pheripheral edge and each of said confronting faces, whereby rotation of the shaft of the rotor assembly oscillates the peripheral edge of the rotor axially back and forth along the confronting faces of said first and second spaced-apart pole pieces with the dimension of said air gap being constant for oscillating the magnetic field in opposite directions axially relative to and through the power windings in the slots of the first and second pole pieces to thereby effect a rate of change of the magnetic field through the power windings without reversing the direction in which the flux flows from one of said first and second pole pieces to the other of said first and second pole pieces and through the rotor and power windings in said first and second pole pieces, to produce alternating voltage in the power windings.

2. The electrical generation device of claim 1 in which the rotor includes a stack of laminations mounted on the shaft at a predetermined angle with respect to the longitudinal axis of said shaft.

3. The electrical generation device of claim 2 in which the lamination stack has an elliptical surface and circular edge contour, rotation of the lamination stack by rotation of the shaft generating a cylindrical surface of revolution.

4. The electrical generation device of claim 2 in which the lamination stack provides a conducting medium for the magnetic flux between the spaced-apart pole pieces, the interactions between the magnetic flux and the power windings producing forces upon the lamination stack in response to its cam-like action and upon the shaft assembly which imparts a torque to the rotor assembly.

5. The electrical generation device of claim 2 in which the lamination stack is in energy resonance with the power windings and a prime mover when connected to the shaft of the rotary assembly which provides a bi-directional power flow between the prime mover and the electrical generation device.

6. The electrical generation device of claim 1 in which the predetermined locations of the slots and the disposition of power windings therein are adapted to be in energy resonance with the rotor stack.

7. The electrical generation device of claim 1 in which the rate of change of the magnetic field with respect to the power winding induces a sixty-cycle voltage within the power windings when a rotational speed of eighteen hundred RPM is imposed upon the shaft of the rotor assembly.

8. The electrical generator device of claim 1 in which the external magnetic field circuit has a plurality of spaced-apart pole pieces.

9. The electrical generation device of claim 1 in which the external magnetic circuit has a pair of spaced-apart pole pieces, the rotor assembly in response to the oblique orientation of the rotor thereof producing two cycles of alternating voltage per revolution of the rotor assembly.

10. An alternator comprising:
a hollow cylindrical stator with the longitudinal axis thereof extending along a predetermined central axis, the stator having a pair of pole pieces each of which extends radially inwardly toward one another with the inner end of each pole piece having a face portion spaced apart from the other, the stator having means for providing a magnetic field extending through the stator and in a direction from one pole piece to the other for establishing a single magnetic polarity in each of said pole pieces, with the polarity of the pole pieces being opposite to each other, a power winding disposed adjacent to the face portion of each pole piece with the turns of each winding substantially facing the face portion of the pole piece; and
a rotor of permeable material disposed within the hollow cylindrical stator and extending along the length thereof between the face portions of the pole pieces, the rotor being mounted for rotation about the central axis of the stator and being adapted to be driven in rotation, the rotor having its oppositely disposed end portions extending in spaced apart planes disposed at an oblique angle to the axis of rotation of the rotor assembly and substantially parallel to one another, said rotor having a peripheral edge extending about said axis of rotation, said peripheral edge being substantially cylindrical and nested within the cylindrical stator said peripheral edge having an axial dimension smaller than the axial dimension of the face portions of the pole pieces, segments of the peripheral edge of the rotor being adjacent to face portions of the pole pieces throughout a rotation of the rotor, when the rotor is driven in rotation the rotor being adapted to couple therethrough the magnetic field in a direction from one pole piece to the other with each segment of the peripheral edge of the rotor as said peripheral edge of the rotor is reciprocating in opposite axial directions with respect to and adjacent the face portion of each pole piece and the

power windings thereof with a substantially harmonic motion and with a substantially constant air gap, in axial section, between each face portion and the peripheral edge of the rotor, said harmonic motion of the rotor transporting the magnetic field axially back and forth across each power winding without reversing the direction in which flux flows from one pole piece to the other and through the rotor and power windings adjacent said pole pieces, thereby inducing an alternating voltage in the power windings.

11. An alternator in accordance with claim 10 in which the face portion of each pole piece is a portion of a cylinder having its central longitudinal axis disposed substantially along the central axis of the stator.

12. An alternator in accordance with claim 10 in which the turns of each power winding have portions extending spaced apart from one another and at an angle to the central axis of the stator and additional portions extending spaced apart from one another in a direction substantially parallel to the central axis of the stator.

13. An alternator in accordance with claim 10 in which the turns of each power winding have portions extending spaced apart from one another and substantially at right angles to the central axis of the stator and additional portions extending spaced apart from one another and substantially parallel to the central axis of the stator.

14. An alternator in accordance with claim 10 in which the face portion of each pole piece contains a recess corresponding to the form of the turns of the power winding therein and being adapted to receive the turns of the power winding therein.

15. An alternator in accordance with claim 14 in which the recess is in the form of a slot extending into the face portions of the pole pieces.

16. An alternator in accordance with claim 10 in which the means for providing a magnetic field extending through the stator and the rotor comprises a field winding having turns disposed about the stator and adapted to produce a magnetic field therein.

17. An alternator in accordance with claim 10 in which each pole piece is a portion of a cylinder having its central longitudinal axis disposed substantially along the central axis of the stator and in which the peripheral of the rotor when rotating generates a cylindrical figure of revolution which has a radius extending perpendicular to the face of each pole piece to form a narrow air gap therewith.

18. An alternator in accordance with claim 10 in which the rotor comprises a plurality of laminations to reduce eddy currents therein the plurality of laminations extending substantially parallel to the oblique angle of the end portions of the rotor.

19. An alternator in accordance with claim 15 in which the oblique angle of the plane of the end portions of the rotor to the pivotal axis thereof is approximately forty-five degrees.

20. An alternator in accordance with claim 10 in which the turns of each power winding include portions oppositely disposed at a predetermined distance along the pole piece in the direction of the central axis of the stator and extending with respect thereto at an angle to the central axis of the stator and in which a portion of the periphery of the rotor reciprocates in opposite directions relative thereto along a path of travel at least equal to and in alignment with the predetermined distance between the portions of the turns.

21. An alternating current motor comprising:
a rotor assembly having an axially extending, rotatable shaft, a rotor affixed obliquely to the shaft and including an outer peripheral edge having an axial dimension provided between axially spaced-apart end faces of said rotor, said end faces being obliquely oriented relative to the shaft, circumferentially spaced segments of said outer peripheral edge being in different axial positions along the shaft as a result of the oblique orientation of the rotor to the shaft;
an external magnetic field circuit having at least first and second spaced-apart pole pieces, means for providing a magnetic field extending through the external magnetic field circuit for establishing a single magnetic polarity in each pole piece, with the magnetic poles of said at least first and second spaced-apart pole pieces being of opposite polarities, each of the pole pieces being of an axial dimension greater than the axial dimension of the outer peripheral edge of the rotor and having a face with a plurality of slots therein, and a plurality of power windings disposed in the slots, said power windings being adapted to be connected to a source of alternating current; and
means for mounting the rotor assembly for relative motion with respect to the first and second pole pieces of the external magnetic field circuit by rotatably mounting the shaft relative to the external magnetic field circuit such that circumferentially spaced segments of the peripheral edge of the rotor, in axial section, are substantially parallel to and spaced from confronting faces of said first and second spaced-apart pole pieces for defining an air gap between said peripheral edge and each of said confronting faces, whereby the power windings when energized by the alternating current source and linked by the magnetic field extending through the external magnetic circuit producing a torque which rotatably drives the rotor assembly to cause the peripheral edge of the rotor to oscillate back and forth axially relative to the confronting faces of the first and second spaced-apart pole pieces with the dimension of the air gap being substantially constant and without changing the direction of flux flow from one of said first and second pole pieces to the other of said first and second pole pieces and through the rotor and power windings in said first and second pole pieces.

22. The alternating current motor of claim 21 in which the rotor includes a lamination stack mounted on the shaft at a predetermined angle with respect to the longitudingal axis of said shaft.

23. The alternating current motor of claim 22 in which the lamination stack provides a conducting medium for the magnetic flux between the spaced apart pole pieces and the application of an alternating current to the power windings causes a harmonic motion of the external magnetic field with respect to the spaced apart pole pieces, giving rise to lateral forces, in accordance with the Lenz law, which are brought to bear upon the cam-like structure of the rotor, thereby developing a torque which causes it to rotate.

24. An alternating current motor comprising:
a hollow cylindrical stator with the longitudinal axis thereof extending along a predetermined central

axis, the stator having a pair of pole pieces each of which extends radially inwardly toward one another with the inner end of each pole piece having a face portion spaced apart from the other, the stator having means for providing a magnetic field extending through the stator and from one pole piece to the other for establishing a single magnetic polarity in each of said pole pieces, with the polarity of the pole pieces being opposite to each other, a power winding disposed adjacent to the face portion of each pole piece with the turns of each winding substantially facing the face portion of the pole piece, the power winding being adapted to be connected to a source of alternating current; and

a rotor of permeable material disposed within the hollow cylindrical stator and extending along the length thereof between the face portions of the pole pieces, the rotor being mounted for rotation about the central axis of the stator, the rotor having its oppositely disposed end portions extending in spaced apart planes disposed at an oblique angle to the axis of the stator and substantially parallel to one another, said rotor having a peripheral edge extending about the axis, said peripheral edge being substantially cylindrical and nested within the cylindrical stator, said peripheral edge having an axial dimension smaller than the axial dimension of the face portions of the pole pieces, segments of the peripheral edge of the rotor being adjacent to face portions of the pole pieces with a constant air gap, in axial section, throughout a rotation of the rotor, the rotor being adapted to couple therethrough the magnetic field in a direction from one pole piece to the other with each segment of the peripheral edge of the rotor as said peripheral edge of the rotor is reciprocating in opposite axial directions with respect to and adjacent the face portion of each pole piece and the power winding thereof with a substantially harmonic motion produced by the interaction of the external magnetic field and the magnetic field developed by the presence of an alternating current within the power windings of each pole piece, and giving rise to lateral forces, in accordance with the Lenz law, which are brought to bare upon the cam-like structure of the rotor, thereby developing a torque which causes said rotor to rotate without changing said direction of flux flow from one pole piece to the other and through the rotor and power windings adjacent said pole pieces.

25. An alternating current motor in accordance with claim 24 in which the turns of each power winding have portions extending spaced apart from one another and at an angle to the central axis of the stator and additional portions extending spaced apart from one another in a direction substantially parallel to the central axis of the stator.

26. An alternating current motor in accordance with claim 24 in which the means for providing a magnetic field extending through the stator and the rotor comprises a field winding having turns disposed about the stator and adapted to produce a magnetic field therein.

27. An alternating current motor in accordance with claim 24 in which each pole piece is a portion of a cylinder having its central longitudinal axis disposed substantially along the central axis of the stator and in which the periphery of the rotor when rotating generates a cylindrical figure of revolution which has a radius extending perpendicular to the face of each pole piece to form a narrow air gap therewith.

28. An alternating current motor in accordance with claim 24 in which the rotor comprises a plurality of laminations to reduce eddy currents therein, the plurality of laminations extending substantially parallel to the oblique angle of the end portions of the rotor.

29. An alternating current motor in accordance with claim 28 in which the oblique angle of the plane of the end portions of the rotor to the pivotal axis thereof is approximately forty-five degrees.

30. An alternating current motor in accordance with claim 24 in which the turns of each power winding include portions oppositely disposed at a predetermined distance along the pole piece in the direction of the central axis of the stator and extending with respect thereto at an angle to the central axis of the stator and in which a portion of the periphery of the rotor reciprocates in opposite directions relative thereto along a path of travel at least equal to and in alignment with the predetermined distance between the portions of the turns.

31. The electrical generation device of claim 1 wherein the windings in the slots are free of the air gaps.

32. The alternating current motor of claim 21 wherein the windings in the slots are free of the air gaps.

* * * * *

Dynafux Alternator Patent Application
Multi-Pole Switched Reluctance D.C. Motor With A Constant Air Gap And Recovery Of Inductive Field Energy
US2013/0187586 A1

This is a copy of the latest Dynaflux Alternator Patent Applications.

THE MEANING OF UNITY IN ENERGY CONVERSION SYSTEMS

(19) **United States**
(12) **Patent Application Publication**
Murray, III

(10) Pub. No.: US 2013/0187586 A1
(43) Pub. Date: Jul. 25, 2013

(54) MULTI-POLE SWITCHED RELUCTANCE D.C. MOTOR WITH A CONSTANT AIR GAP AND RECOVERY OF INDUCTIVE FIELD ENERGY

(76) Inventor: **James F. Murray, III**, Oklahoma City, OK (US)

(21) Appl. No.: **13/562,233**

(22) Filed: **Jul. 30, 2012**

Related U.S. Application Data

(63) Continuation-in-part of application No. 12/993,941, filed on Dec. 3, 2010, now Pat. No. 8,373,328, filed as application No. PCT/US09/46246 on Jun. 4, 2009, Continuation-in-part of application No. 13/390,437, filed on Feb. 14, 2012, filed as application No. PCT/US10/45298 on Aug. 12, 2010.

(60) Provisional application No. 61/058,824, filed on Jun. 4, 2008, provisional application No. 61/234,011, filed on Aug. 14, 2009.

Publication Classification

(51) Int. Cl.
H02P 6/14 (2006.01)
H02K 1/14 (2006.01)
H02K 19/06 (2006.01)

(52) U.S. Cl.
CPC *H02P 6/14* (2013.01); *H02K 19/06* (2013.01); *H02K 1/14* (2013.01)
USPC **318/400.37**; 310/46; 310/216.006

(57) **ABSTRACT**

A Back EMF reducing DC motor system and method of operation are disclosed. The disclosed system and method are designed to exploit Transformer Voltage properties and include a rotor element shaped to periodically move a flux zone along a stator face. Incoming DC motor power from an external source may be appropriately conditioned and applied to a power supply. Storage Capacitors may also communicate with the power supply. A controller receives power from the power supply and communicates with the DC motor. A position sensor or other indicator may also communicate DC motor operational conditions to the controller. A recapture storage device may receive recaptured power from the DC motor via the controller. The recaptured power may be used to power an external load, or to reduce the input power necessary to operate the DC motor.

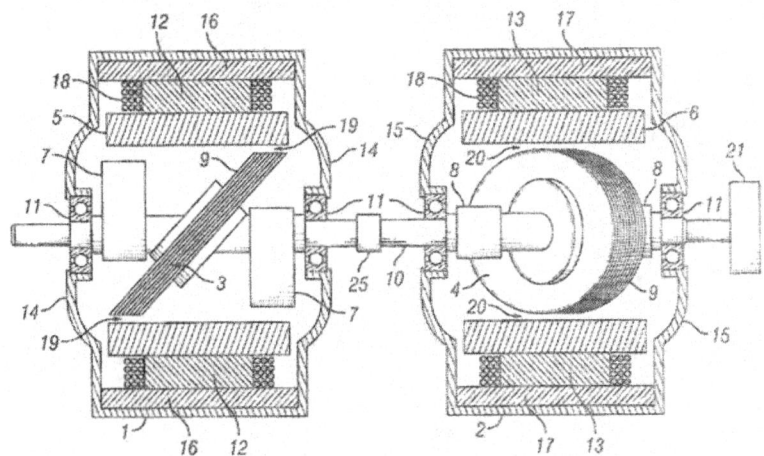

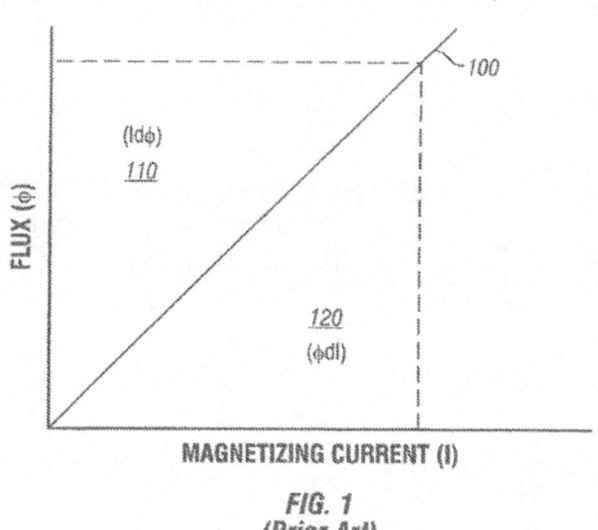

FIG. 1
(Prior Art)

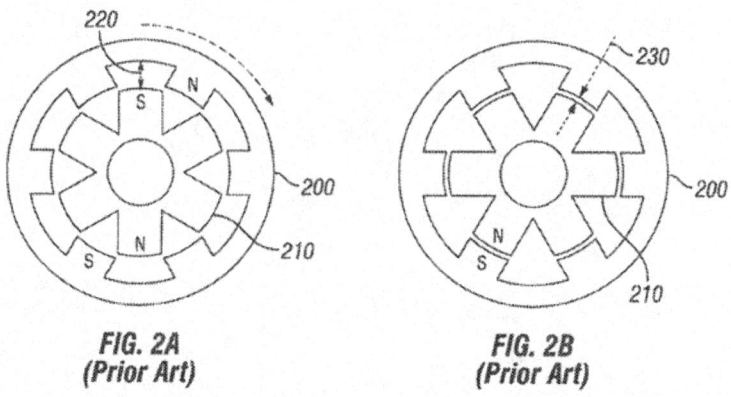

FIG. 2A
(Prior Art)

FIG. 2B
(Prior Art)

THE MEANING OF UNITY IN ENERGY CONVERSION SYSTEMS

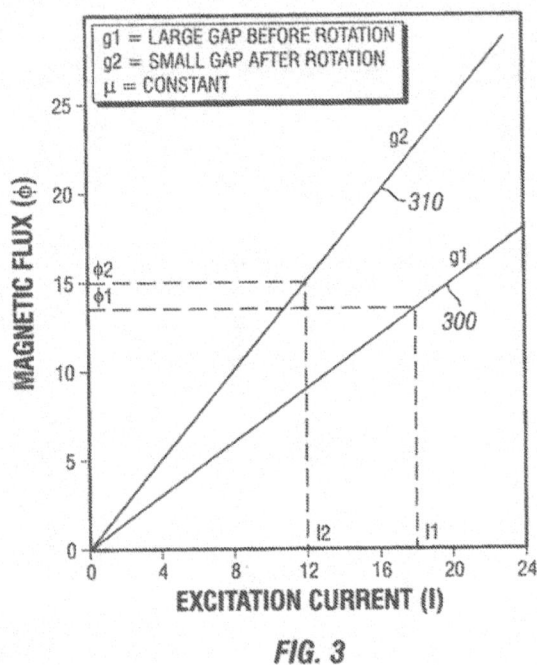

FIG. 3

THE MEANING OF UNITY IN ENERGY CONVERSION SYSTEMS

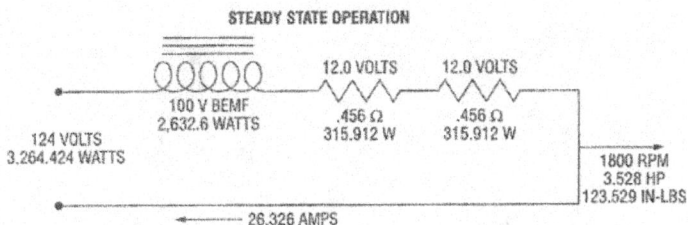

FIG. 4A

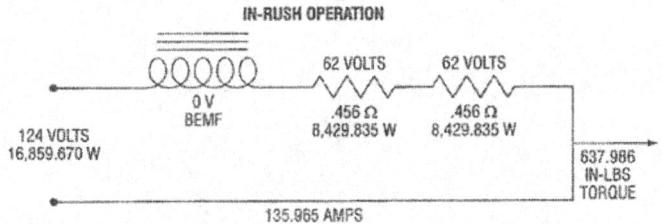

FIG. 4B

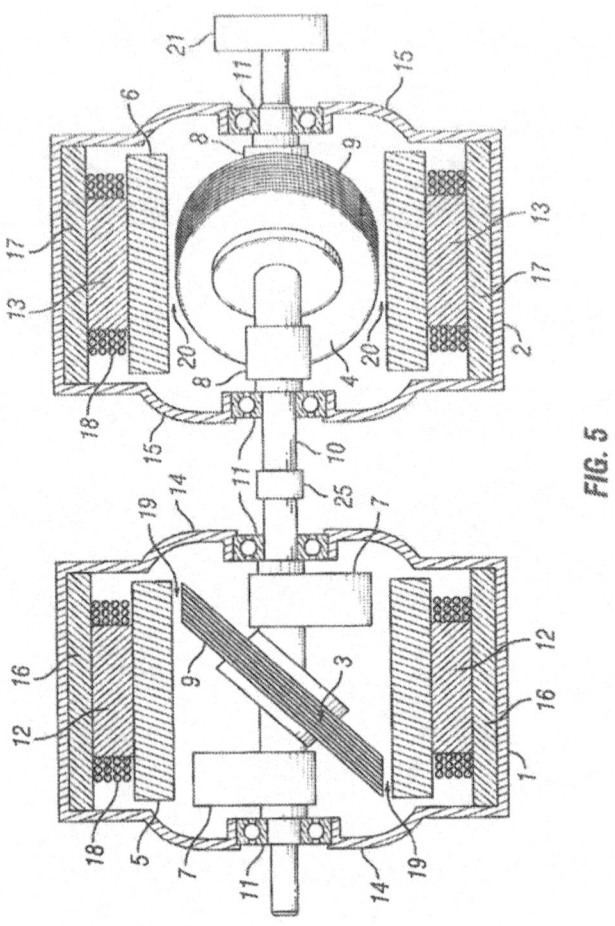

FIG. 5

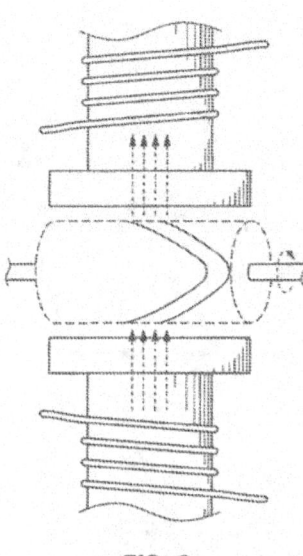

FIG. 6

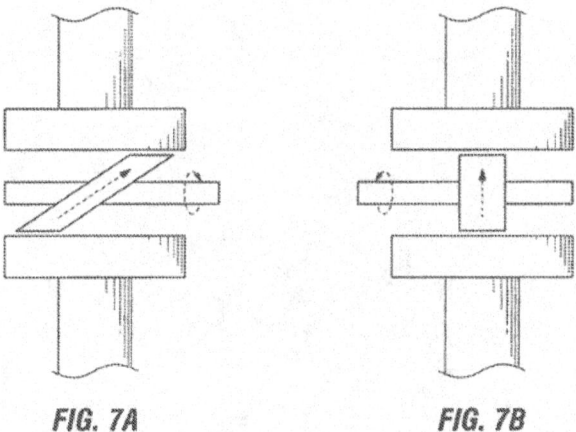

FIG. 7A **FIG. 7B**

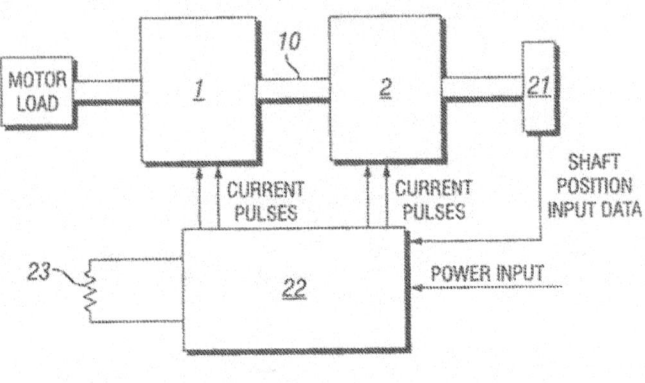

FIG. 8

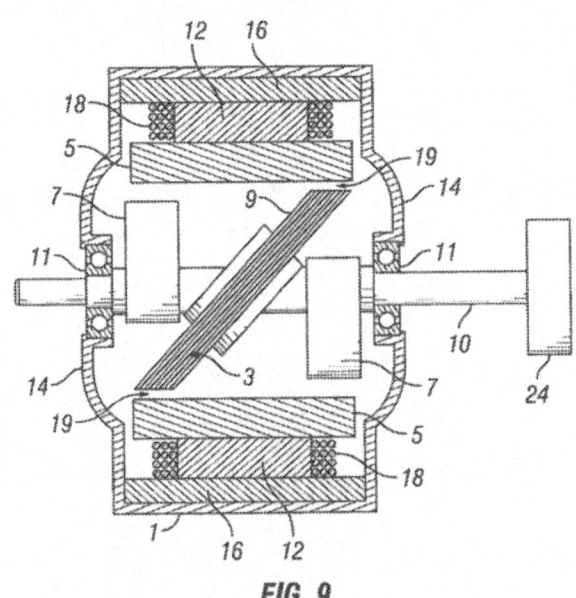

FIG. 9

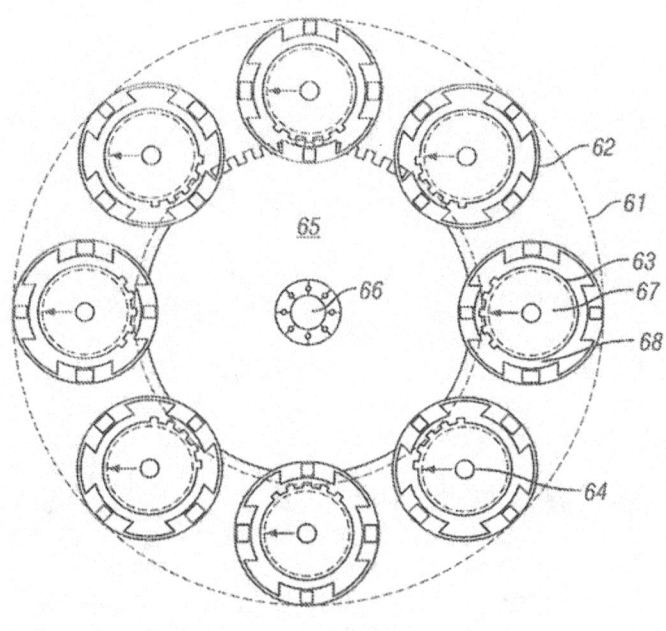

FIG. 10A

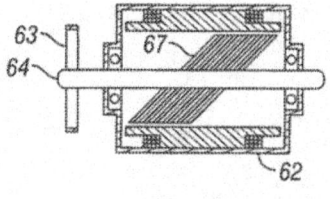

FIG. 10B

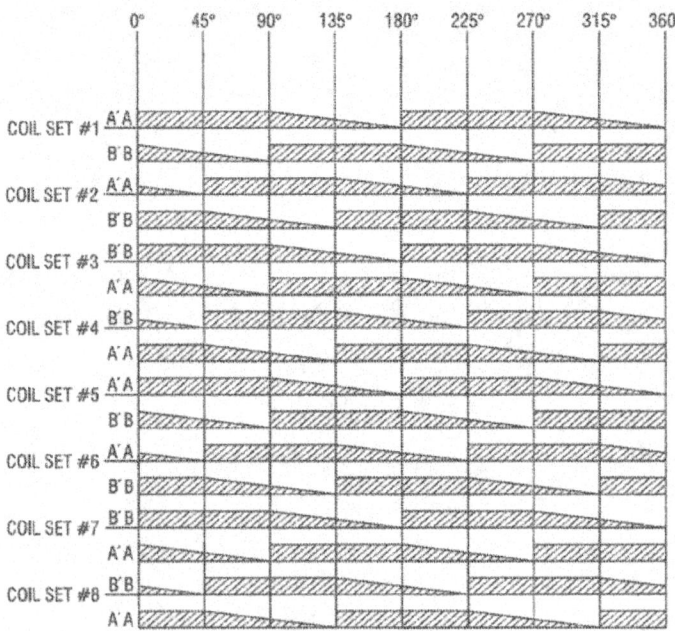

FIG. 11

THE MEANING OF UNITY IN ENERGY CONVERSION SYSTEMS

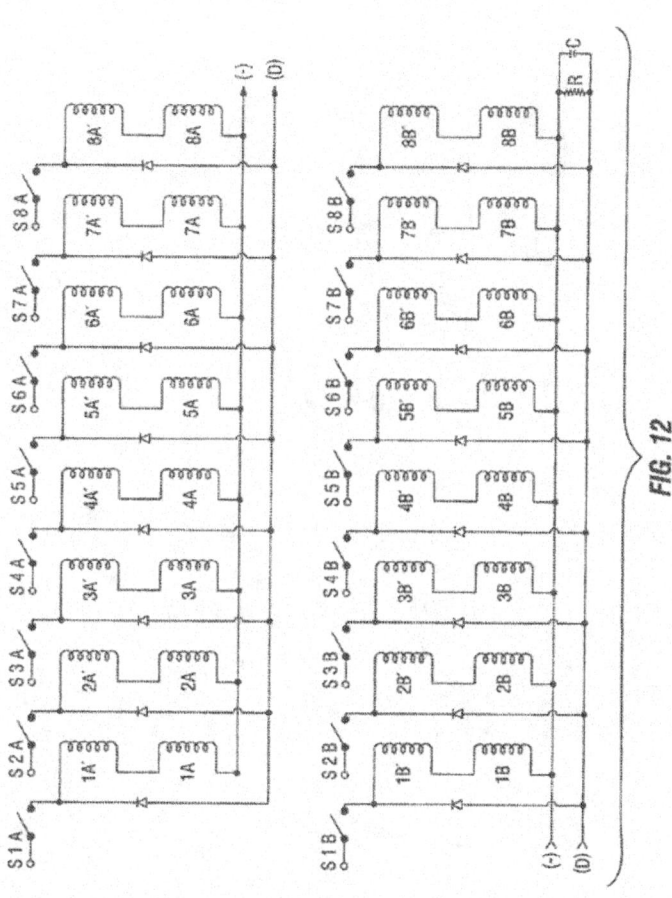

FIG. 12

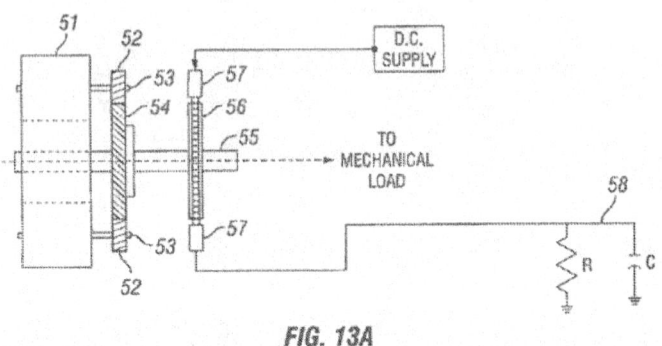

FIG. 13A

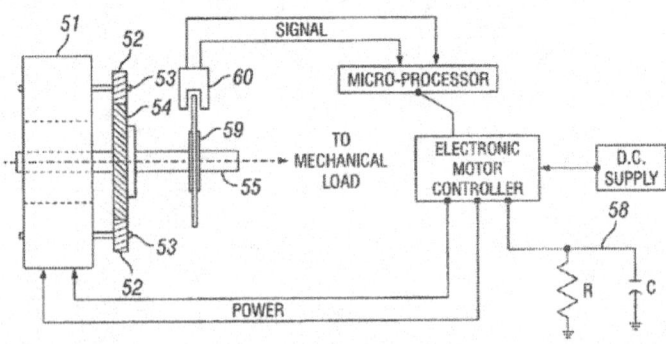

FIG. 13B

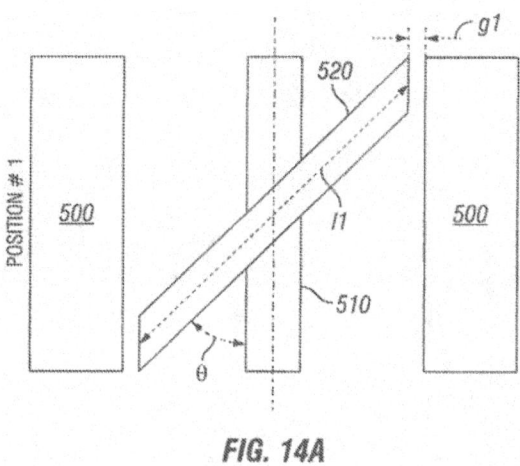

FIG. 14A

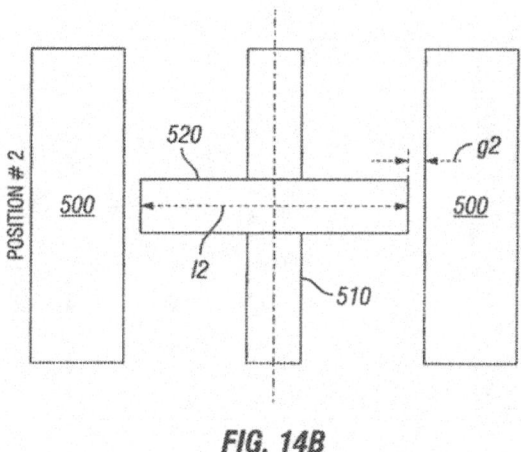

FIG. 14B

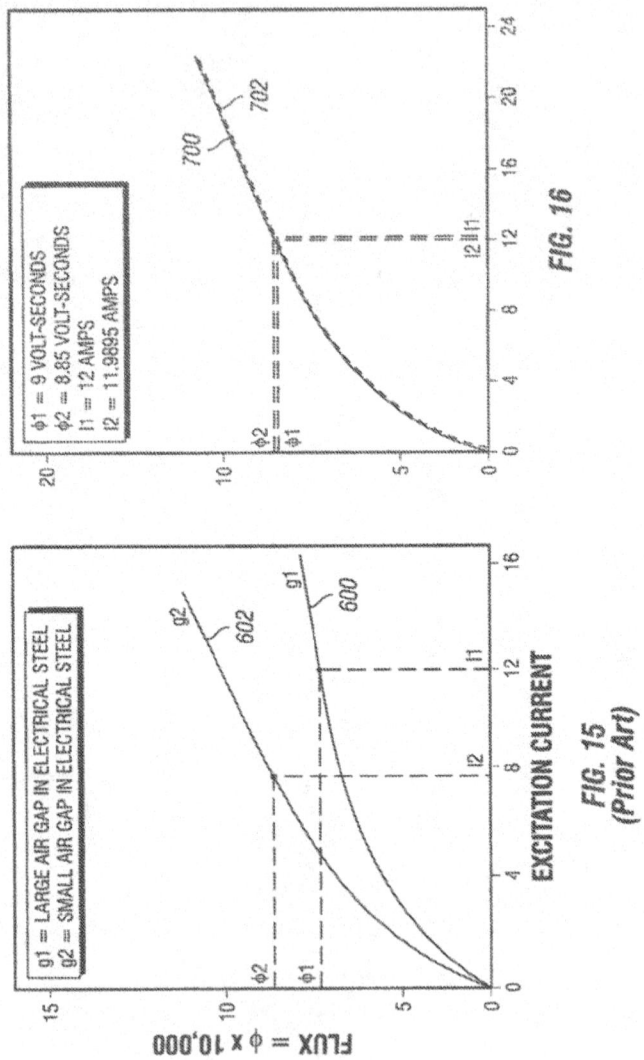

FIG. 16

FIG. 15
(Prior Art)

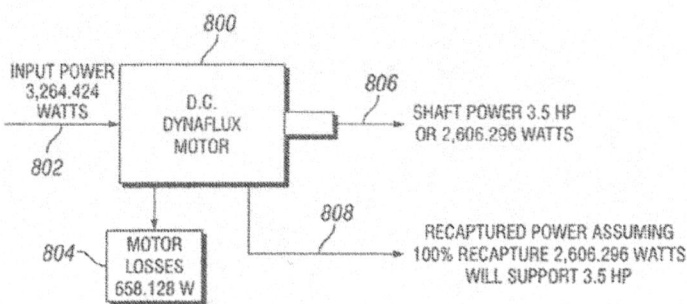

FIG. 17A

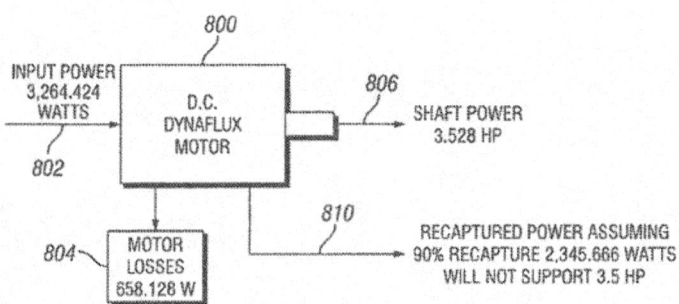

FIG. 17B

THE MEANING OF UNITY IN ENERGY CONVERSION SYSTEMS

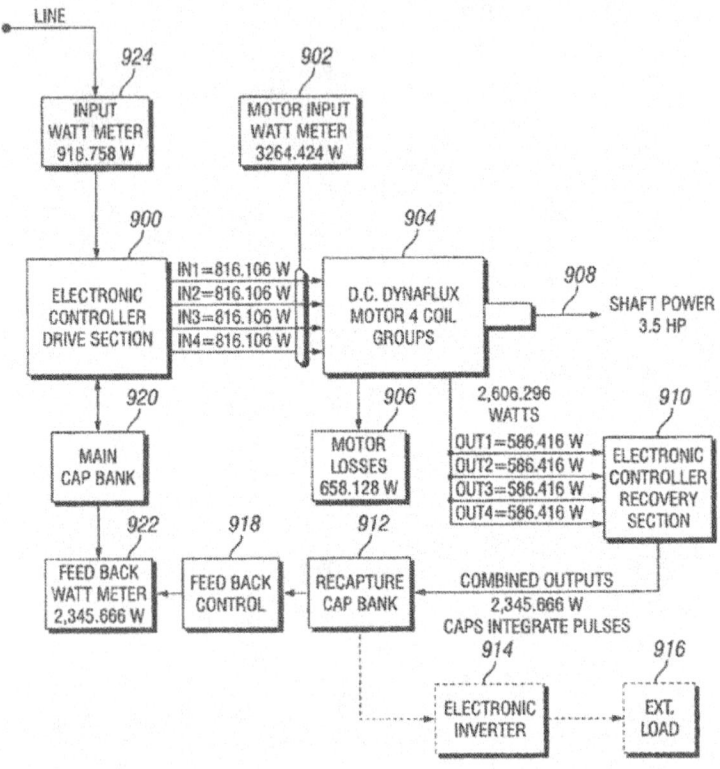

FIG. 18

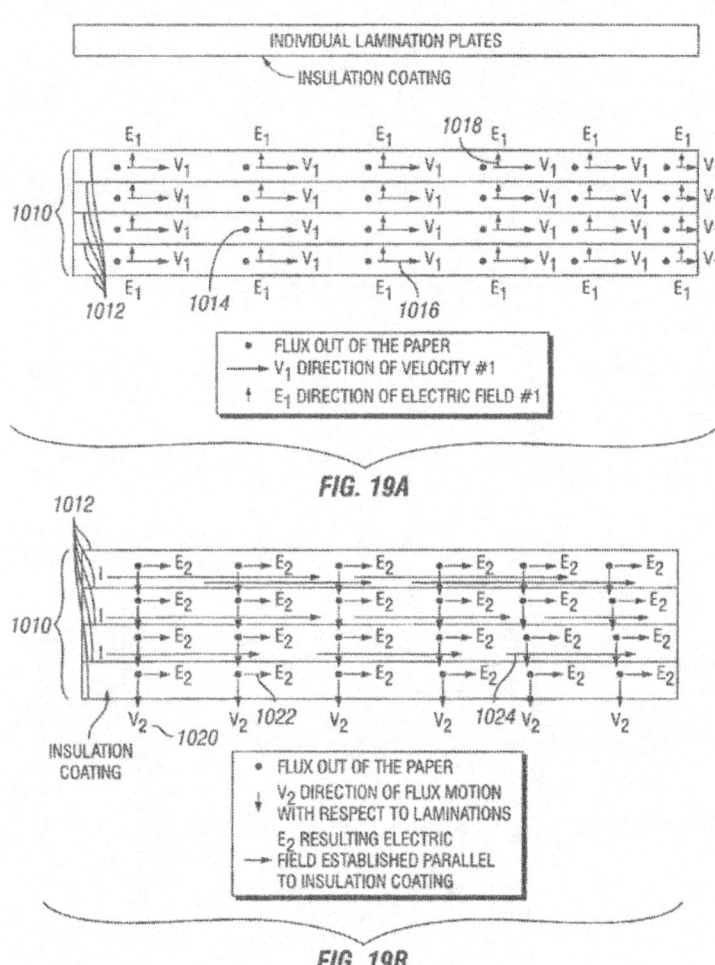

FIG. 19A

FIG. 19B

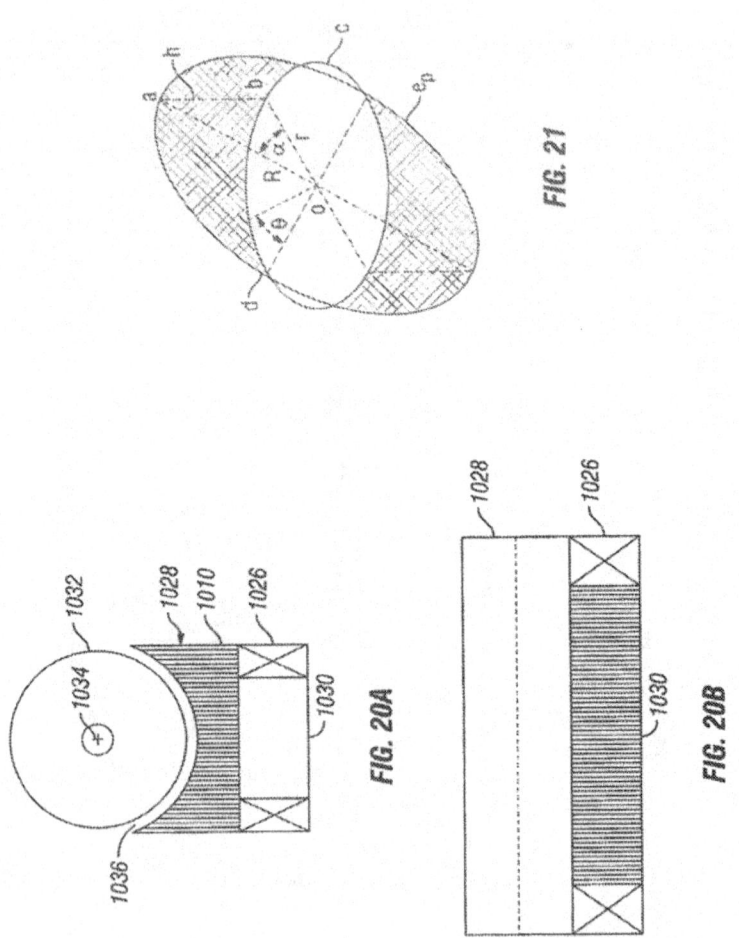

FIG. 21

FIG. 20A

FIG. 20B

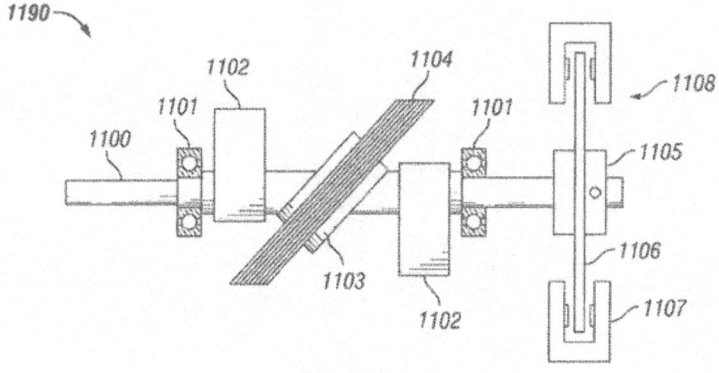

FIG. 22

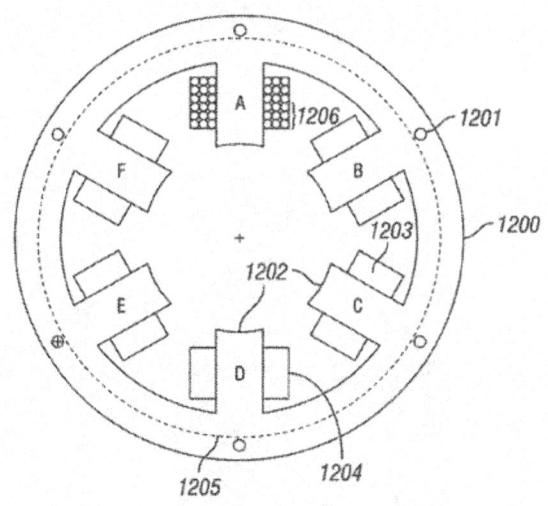

FIG. 23

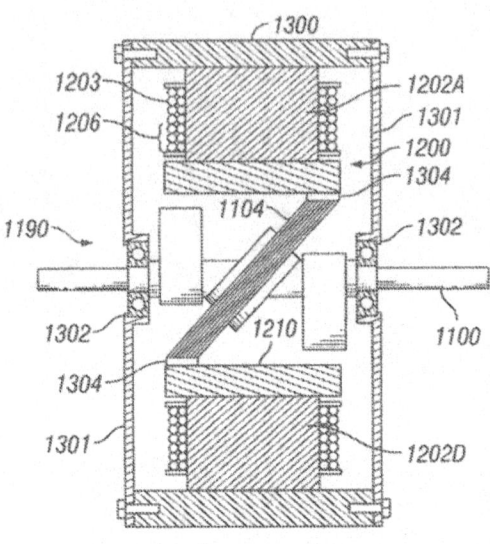

FIG. 24

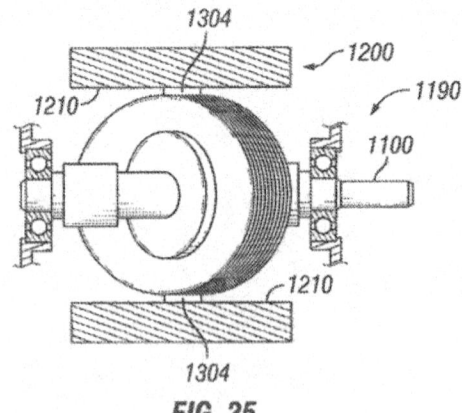

FIG. 25

THE MEANING OF UNITY IN ENERGY CONVERSION SYSTEMS

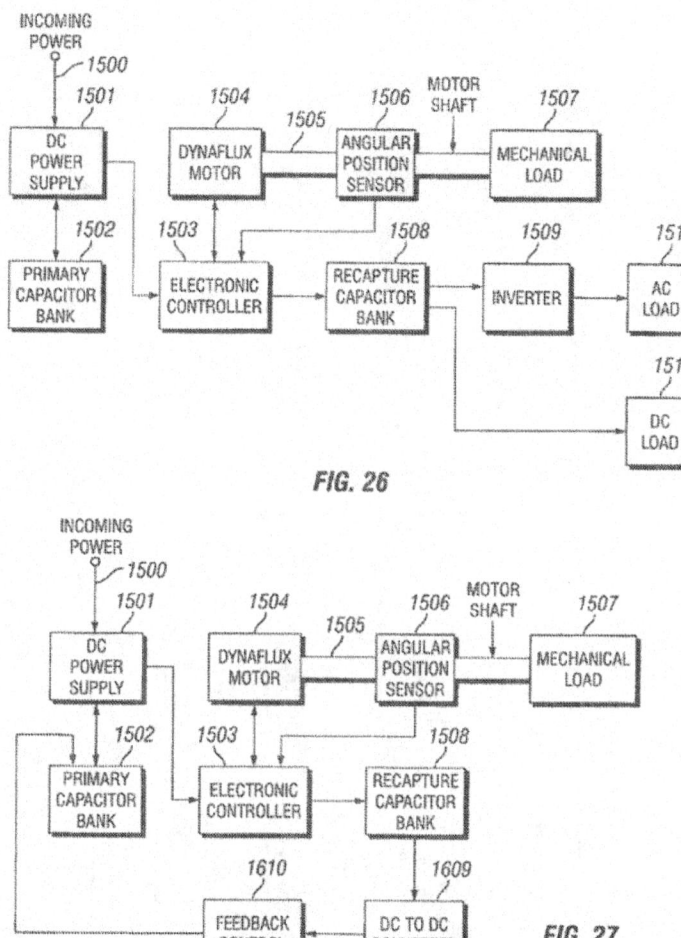

FIG. 26

FIG. 27

THE MEANING OF UNITY IN ENERGY CONVERSION SYSTEMS

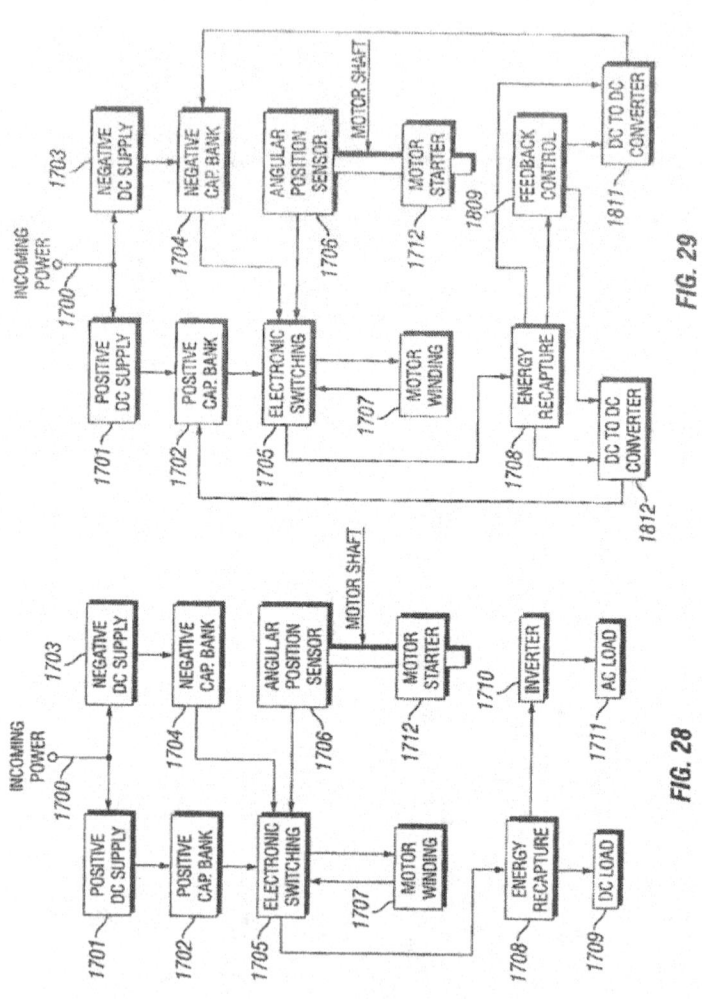

FIG. 29

FIG. 28

THE MEANING OF UNITY IN ENERGY CONVERSION SYSTEMS

US 2013/0187586 A1　　　　　　　　　　　　　　　　　　　　　　　　　　　　　Jul. 25, 2013

MULTI-POLE SWITCHED RELUCTANCE D.C. MOTOR WITH A CONSTANT AIR GAP AND RECOVERY OF INDUCTIVE FIELD ENERGY

CROSS REFERENCE TO RELATED APPLICATIONS

[0001] This application is a continuation-in-part of application Ser. No. 12/993,941, which has a 35 U.S.C. §371(c) date of Dec. 3, 2010, and which in turn is a 35 U.S.C. §371 filing of Application No. PCT/US09/46246, filed Jun. 4, 2009, which in turn claims the benefit under 35 U.S.C. §119 to provisional Application No. 61/085,824, filed Jun. 4, 2008, and the entire contents of each are hereby incorporated by reference.

[0002] This application is also a continuation-in-part of application Ser. No. 13/390,437, which has a 35 U.S.C. §371 (c) date of Feb. 14, 2012, and which in turn is a 35 U.S.C. §371 filing of Application No. PCT/US10/45298, filed Aug. 12, 2010, which in turn claims the benefit under 35 U.S.C. §119 to provisional Application No. 61/234,011, filed Aug. 14, 2009.

[0003] This application is also related to the following concurrently-filed Applications: application Ser. No. _____, titled "Controller for Back EMF Reducing Motor;" application Ser. No. _____, titled "Three Phase Synchronous Reluctance Motor With Constant Air Gap And Recovery Of Inductive Field Energy;" and Provisional Application No. _____, titled "Multi-pole Electrodynamic Machine With A Constant Air Gap And An Elliptical Swash-Plate Rotor To Reduce Back Torgye;" each of which are hereby incorporated by reference.

FIELD OF THE INVENTION

[0004] The disclosed inventions relate to the field of direct energy conversion and the production of mechanical torque from the utilization of an electric current, and to the field of electric motors and to utilization of direct current as a "motive force." The disclosed inventions also relate to the field of power conversion devices which transform electrical power into rotary mechanical power.

[0005] Some disclosed embodiments relate to a class of motor having multiple stator and rotor sections, such that each rotor section is associated with a specific stator section, although attached to a single output shaft. The lateral axis of each rotor section may be disposed at an oblique angle with respect to the axis of the common shaft, and angularly displaced in accordance with the number of rotor sections employed, for example: 90 mechanical degrees for two rotors, 120 degrees for three rotors, etc.

[0006] Some disclosed embodiments also relate to multiple motors having two or more motor sections, operating in parallel, each of which is comprised of a stator having two or more salient poles, and a rotor geometry devoid of coils or windings of any kind, affixed obliquely to a motor output shaft, and so disposed as to ensure a constant air gap between the rotor body and the salient poles of an associated stator section.

[0007] Some embodiments of the invention also relate to multiple motor sections with their associated armatures, mechanically positioned out of phase with one another, but mounted so as to allow the output pinions of each individual motor to impinge upon a common output gear, of larger diameter, mounted upon a separate but common output shaft, such that each individual motor's output is combined mechanically, and afforded an amplification of torque.

[0008] Some embodiments of the invention also relate to a single motor having a stator section with salient poles, and a rotor geometry devoid of windings, affixed obliquely to a motor output shaft, and disposed as to ensure a constant air gap between the rotor body and the salient poles of the stator section.

[0009] Some embodiments of the invention relate to a switched reluctance D.C. motor motor having a stator section with salient poles, and a rotor geometry devoid of windings, affixed obliquely to a motor output shaft, and disposed as to ensure a constant air gap between the rotor body and the salient poles of the stator section

BACKGROUND

[0010] Notwithstanding the increased interest in energy conversion over the recent decades, no substantial advances have been made in increasing the conversion efficiency of electric motors. Rather, the art has made incremental advances relating to improved magnetic materials, more powerful permanent magnets, and sophisticated electronic switching devices. Such improvements, at best, relate to very small increases in overall efficiency, usually gained at very considerable expense.

[0011] Patents in this area include: U.S. Pat. Nos. 2,917, 699; 3,132,269; 3,321,652; 3,956,649; 3,571,639; 3,398,386; 3,760,205; 4,639,626 and 4,659,953. Also in this area are EPO patent no. 0174290 (March 1986); German patent no. 1538242 (October 1969); French patent no. 2386181 (October 1978) and UK patent no. 1263176 (211972).

[0012] The basic concept employed in earlier motor art is the interaction between a current carrying conductor(s) and a magnetic field of some kind. This fact is true regardless of motor type. This basic concept appears in DC motors, single phase AC motors, poly phase induction slip motors, which utilize a rotating magnetic field, and in poly phase synchronous Motors with externally excited electromagnetic cores, or permanent magnet cores as the case may be.

[0013] Other types of designs may be found, for example, in the design of stepper motors, which utilize a magnetic "ratcheting" action upon magnetic material in the armature, in response to applied pulses of current, and various types of reluctance motors in which the rotor moves with respect to a salient pole piece, experiencing a large variation in air gap during its motion. But, these devices typically do not have a constant and continuous air gap of fixed dimension between the rotor and the stator.

[0014] The prior art has not produced a multiple phase, multiply segmented stator with individual, obliquely disposed, laminated armatures devoted to each stator section, such that each stator/rotor combination employs a continuous air gap of constant dimension, regardless of the elliptical profile of said armatures, but not employing any current carrying conductors, coils, windings or bars within or upon the armatures, as a means of producing torque upon the output shaft.

[0015] Nor can it be said that the prior art has arranged such motors to cooperate in "parallel fashion," through a reduction gear arrangement so as to provide an amplification of torque while sharing the mechanical load.

[0016] A previous example exists, which describes an alternator having a single rotor canted at an angle, and makes use of the unique rotor design featured within this disclosure. Said rotor was introduced in the power conversion device entitled "Alternator Having Improved Efficiency," which was invented by James F. Murray III, filed as application Ser. No. 07/112,025, on Oct. 21, 1987, and later granted U.S. Pat. No. 4,780,632 on Oct. 25, 1988, and is herein incorporated by reference.

[0017] There are marked differences between the presently disclosed inventions and the inventions disclosed in the "Alternator Having Improved Efficiency," patent ("the Alternator Patent"). A few non-limiting examples of which are listed as follows:

[0018] 1.) Alternator of the Alternator Patent can be operated as a motor only when used in conjunction with the basic motor concepts described herein (i.e., requires field flux and current-carrying conductors).

[0019] 2.) Alternator of the Alternator Patent does not require salient pole projections in order to operate.

[0020] 3.) Alternator of the Alternator Patent makes use of an electromagnetic field winding, or a permanent magnet as its source of magnetic flux.

[0021] 4.) Alternator of the Alternator Patent does not require a shaft position indicator, or a commutator of any kind in order to function.

[0022] 5.) Alternator of the Alternator Patent does not require a position sensitive, electronically controlled, pulsed power supply, in order to generate electricity.

[0023] Other similarities between the Alternator Patent and the presently disclosed inventions include elements possessed by most rotating power converters, such as bearings, shafts, end bells, laminations, mechanical housing, etc.

[0024] As evident from the above discussion, electric motors have been in use for well over 100 years, and they exist in several forms. While, the basic concept has not substantially changed, the manner in which the switching of supply current is controlled has evolved. However, existing motors typically experience performance limitations due to the manner in which Back EMF and inductive field energy are treated. The generation of Back EMF in motors of all kinds is chiefly due to two things: the movement of conductors through a magnetic field, called Speed Voltage, and the rate of change of current through a winding, called Transformer Voltage. Conventional wisdom suggests that Speed Voltage Back EMF is totally unavoidable, and in fact, is necessary for the transformation of electrical power into mechanical power in a typical motor. However, one drawback of Speed Related Back EMF is its parasitic nature that serves to degrade the potential supplied to the motor from an outside source (i.e., the source voltage).

[0025] The parasitic nature of Back EMF arises from, among other things, the mistaken assumption that Back EMF is required to produce torque. This, in turn, leads to design compromises which must be made in order to implement traditional electrodynamic machine geometries. Consider, for example, a conventional DC Motor consisting of a stator with salient field poles, and a rotor-armature with a self-contained commutator. Application of a DC current to the rotor leads produces a rotary motion of the rotor (i.e., motor action). However, the rotation of the rotor conductors in a magnetic field also induces a voltage in the conductor that opposes the current applied to the rotor leads i.e., generator action). These facts actually demonstrate an important aspect of conventional machines; if standard design parameters are always followed, then any motor must perform as a generator while it is running, and any generator must perform as a motor while it is in operation. The explanation of this similarity is because both machines are dependent upon the same basic geometry for their functionality, and so, both motor and generator action occur simultaneously in both devices.

[0026] The above-described basic geometry of a conventional Speed Voltage based system results in the production of parasitic Back EMF as follows. In a Speed Voltage based system, the magnetic flux must interact with an electrical current-carrying conductor (e.g., rotor windings), thereby producing a mechanical force that generates a torque to turn the motor shaft (i.e., a motor action). The subsequent motion of the conductors through the magnetic flux produces a relatively high Back EMF (i.e., acts in opposition to the torque producing current) due to the motion of the conductors with respect to the magnetic flux (i.e., a generator action). In order to continue normal operation, and establish electrical equilibrium, any motor that produces a Back EMF having a constant average value, must draw down on the line-potential in order to overcome the effects of this parasitic Back EMF voltage. Thus, this process of source potential degradation due to Back EMF requires the input of considerable energy from the source in the form of a voltage in order to maintain normal operation.

[0027] Another design factor of conventional Speed Voltage dependent machines is that, typically, as the rotor turns from pole to pole the air gap between the rotor and the stator will vary in width (from a smaller gap when the rotor is "facing" a stator pole, to a larger gap when the rotor is "between" stator poles). This change in the air gap results in a change in the magnetic potential energy within the air gap resulting in the Back EMF component described above. These and other significant issues and inefficiencies persist in traditional DC motor designs.

[0028] Before turning to the improvements and advantages of the disclosed inventions, a brief review of some fundamental concepts for electric motor operation is instructive. The basic premise is that the force developed by a current carrying conductor immersed in a magnetic field is described as (equation 1):

$$F = BIl$$

[0029] where, F is the force developed, B is the flux density, l is the conductor length, and I is the current. This simple equation suggests that a current-carrying conductor situated in a magnetic field will experience a force that is directly proportional to the applied current, the flux density and the length of the conductor. This principle underlies the operation of the millions of electric motors spinning every day in locations all over the world.

[0030] The voltage produced by a conductor moving through a magnetic field can be described using (equation 2):

$$V = Blv$$

[0031] where, V is the voltage developed, B is the flux density, l is the conductor length, and v is the tangential velocity of the conductor as it rotates. Accordingly, if a conductor is moved through a magnetic field by an external motive force (e.g., a prime mover), then the voltage produced may give rise to a current in the conductor, and such a device exhibits generator action. Conversely, if a conductor is carrying a current, and thereby moves through a magnetic field under the influence of the current itself, the device exhibits motor action. However, in the act of moving through the field a voltage is produced within the conductor accordance with equation 2, and acts in such a manner as to diminish the applied current responsible for the conductor's motion, and this produced voltage is typically referred to as a Back EMF.

[0032] Examining the actual power present in the system can be accomplished as follows. Mechanical power can be expressed as the product of Force and Velocity. Velocity is therefore missing from the first relationship (equation 1), but it can be included by multiplying both sides of equation 1 by the additional parameter:

$Fv=BIlv$

[0033] The resulting expression now denotes a form of mechanical power expressed as (equation 3),

$Pm=BIlv$

[0034] where, Pm denotes mechanical power.
[0035] In similar fashion, the voltage expression (equation 2) denotes only potential, not power. Electrical power can be expressed as the product of voltage and current. Current is missing from the second relationship (equation 2), but it can also be included by multiplication to both sides of the equation:

$VI=BlvI$

[0036] The resulting expression now denotes a form of electrical power as (equation 4),

$Pe=BlvI$

[0037] Note that BIlv (equation 3) is equal to BlvI (equation 4), and therefore, Pe must be equal to Pm. This analysis is as expected, and holds with current theories that stipulate the applied power is equal to the output power minus the system losses.

[0038] Another important factor to consider is the magnetic flux in a DC motor. The flux, Φ, can be expressed as (equation 5):

$\Phi=LI$

[0039] where L is the inductance and I is the current. Taking the derivative of the flux expression with respect to time, t, yields:

$d\Phi/dt=d(LI)/dt$

[0040] Substituting V for $d\Phi/dt$ gives (equation 6):

$V=LdI/dt+IdL/dt$

[0041] The first term in equation 6 is the product of inductance (L) and the rate of change of current (I) with respect to time (t). This is the previously discussed Transformer Voltage Vt. The second term is the product of the current (I) and the rate of change of Inductance (L) with respect to time (t). This is the previously discussed Speed Voltage Vs. Thus the relationships for each Voltage type is:

[0042] Transformer Voltage (equation 7),

$Vt=L\,dI/dt$, and

[0043] Speed Voltage (equation 8),

$Vs=I\,dL/dt$.

[0044] Expressing Vt and Vs in terms of the energy can be accomplished as follows. The field energy, Pt, due to the Transformer Voltage may be expressed as follows:

$Pt=IVt$.

[0045] Substituting for Pt and Vt gives:

$dE/dt=Id\Phi/dt$.

[0046] Simplifying to (equation 9):

$dE=Id\Phi$.

[0047] Equation 9 expresses the quantity commonly referred to as the reactive energy. The dissipative energy for the system can, likewise, be expressed as follows. Starting from equation 8, $Vs=IdL/dt$, and realizing that $L=\Phi/I$, then $L=\Phi(I^{-1})$, and $dL/dt=\Phi I^{-2}dI/dt$.

[0048] Substituting $(\Phi I^{-2})dI/dt$ for dL/dt gives:

[0049] $Vs=I(-\Phi/I^2)dI/dt$. Multiplying both sides of the equation by I yields an expression for dissipative power, Ps. But, $VsI=dE/dt$, therefore, $Ps=dE/dt=-\Phi dI/dt$, and (equation 10):

$dE=-\Phi dI$.

[0050] Combining equation 9 and equation 10 the total energy in an air-gap is (equation 11):

$E_t=Id\Phi+\Phi dI$.

[0051] The energy relationship described in equation 11 can be further explained with reference to FIG. 1, which depicts a plot of flux (Φ) versus current (I) of the air gap energy components. As shown, the line 100 represents the total magnetic energy given by (equation 12):

$Em=I\Phi$.

[0052] The region 110 above line 100 indicates the ($I\,d\Phi$) reactive energy region and region 120 below line 100 indicates the ($\Phi\,dI$) dissipative energy region.

[0053] The relevance of this energy relationship can be further explained with reference to FIGS. 2A and 2B which show a cross-sectional representation of a prior art reluctance motor. As shown in FIG. 2A, rotor 210 is in a position between two stator 200 poles yielding the motors largest air gap 220 designated as (g1). In normal operation, when the magnetic poles are energized with the proper magnetic polarity, the flux lines thus created will reach across this gap 220 as they are formed, and cause the rotor 210 to rotate to the position depicted in FIG. 2B, thereby reducing the reluctance in the magnetic circuit and reducing the air gap 230 to its smallest dimension designated as (g2). A torque impulse is also created during this motoring action, and the average mechanical work which is delivered on the rotor 210 will be found to be directly equal to the change in energy (ΦdI) within the air gap.

[0054] Referring now to FIG. 3, which is a double graph representing the energy relationship for the prior art motor illustrated in FIGS. 2A and 2B. The plot labeled 300 corresponding to air gap (g1) represents the relationship between the excitation flux and the excitation current at the point in time where the gap dimension is largest (e.g., air gap 220 as depicted in FIG. 2A). Note the larger value of the excitation current (I_1), and the relatively lower value of the associated flux (Φ_1). This is due to the fact that the large air gap has a high value of magnetic reluctance, and therefore requires substantially more current to produce the associated value of flux (Φ_1). This condition changes for the plot labeled 310 (corresponding to air gap g2), because the air gap has been greatly reduced, and much less current (I_2) is required to establish and hold the flux (Φ_2) within the magnetic circuit. Note that the current has reduced to value I_2, and the flux has actually increased to value Φ_2. This may sound like a positive result, but actually, it is not, because this large change in the flux (Φ) is also responsible for the production of an associated Back EMF.

THE MEANING OF UNITY IN ENERGY CONVERSION SYSTEMS

[0055] For illustrative purposes, the following four calculations using equation 11 can be made representing the component energies associated with each air gap size (g1 and g2).

[0056] For a gap size g1: $\Phi_1 dI=(13.5)(18-12)=81$ Joules, and $I_1 d\Phi=(18)(15-13.5)=27$ Joules. For a gap size g2: $\Phi_2 dI=(15)(18-12)=90$ Joules, and $I_2 d\Phi=(12)(15-13.5)=18$ Joules.

[0057] Thus, each energy component has a different value, but much more interesting to note is that the total energies E1 and E2 which represent the energy for air gap sizes of g1 and g2, respectively, are equal (27+81)=(18+90)=108 Joules. This is consistent with the understanding that the motor shaft energy and motor input energy are equal in a motor of standard design, and co-exist within the motor structure. Hence, the term co-energy.

[0058] In further illustration of conventional DC motor operation, consider the following example of normal, Speed Voltage dependent operation. As depicted schematically in FIG. 4A, an exemplary standard DC motor with a power rating of 3.528 Horse Power has the following characteristics:

[0059] Full Load Speed=1800 RPM.
[0060] Continuous Shaft Torque=123.529 in-Lbs.
[0061] Terminal Voltage=124 Volts DC.
[0062] Full Load Current=26.326 amps.
[0063] Copper Losses=315.912 watts.
[0064] Other Losses 315.912 watts in the aggregate.
[0065] Back EMF Power Loss=2632.600 watts.
[0066] Shaft Power=3.528 H.P.
[0067] Total Input Power 3264.424 watts.
[0068] System Efficiency=80.645%.

[0069] Accordingly, if the proper voltage is applied to the motor terminals, and the mechanical load does not vary, the above properties should prevail indefinitely after thermal equilibrium has been reached However, this same example DC motor will have drastically different properties upon first being started. This is illustrated by the diagram in the second diagram in FIG. 4B, showing the start-up, or in-rush operation.

[0070] At the instant illustrated, the DC motor has not yet begun to rotate, and there is no Back EMF, but the starting torque is relatively large at 637.986 in-lbs, which is 5.165 times the running torque. The Back EMF that develops as a function of the motor's increasing rotational speed reduces the start-up current of 135.965 amps down to the full load ampere (FLA) value of 26.326 amps. This "high start-up current," behavior is standard and expected in conventional Speed Voltage dependent motors.

[0071] Bearing these facts in mind, it stands to reason that for two, otherwise-identical, electric motors, the one that employs a larger, or surplus, number of winding turns per pole would experience a comparatively higher inductance L, and correspondingly, a relatively higher total Back EMF, resulting from the sum of Vs and Vt. As a way to avoid this occurrence, it is typical in the prior art of electric motor design that the winding turns per pole are generally kept to a minimum, for a given operational voltage, thus allowing the Speed Voltage component to drive the design criteria, and minimize the Transformer Voltage component.

[0072] However, this engineering trade-off, of keeping inductance L low by using fewer windings, diminishes the amount of stored energy in the motor's magnetic circuit, and causes motor performance to he tied to the characteristics imposed by the Speed Voltage component of the Back EMF, most notably, the requirement for a higher magnitude source voltage and reduced torque output. Other motor design drawbacks and Back EMF issues also exist in prior systems.

SUMMARY

[0073] An electric motor is disclosed, some embodiments having a motor segment having a stator, having stator poles and stator windings and a rotor having a flux path element. For some embodiments, the flux path element is attached to a rotor shaft at an oblique angle to the longitudinal axis of the shaft. The flux path element has a shape that provides a uniform constant air gap between it and the stator poles when the shaft is rotated.

[0074] An electric motor is disclosed, some embodiments having a plurality of motor segments, each segment having a stator, having stator poles and stator windings and a rotor having a flux path element. For some embodiments, the flux path elements are attached to a rotor shaft at an oblique angle to the longitudinal axis of the shaft. The flux path elements have a shape that provides a uniform air gap between them and the stator poles when the shaft is rotated. The rotor shafts of said motor segments are mechanically coupled to each other.

[0075] In an embodiment, the flux path elements comprise a silicon steel lamination stack or a solid ferrite plate. In a further embodiment, the motor has a shaft angle sensor and a motor controller, and the motor controller receives a shaft angle from the sensor and supplies current pulses to the stator windings according to the shaft's angular position signal.

[0076] In a further embodiment, the stator poles are positioned in pole pairs with the rotor and rotor shaft between them and form isolated stator magnetic field circuits when the stator windings are supplied with electrical current, such that a magnetic field is established having a single magnetic polarity in each of the poles of said pole pairs, with each pole of the pole pairs having opposite magnetic polarity. In further embodiments more than two poles are installed in each stator section.

[0077] In a further embodiment, the rotor flux path elements have a shape defined by the volume contained between two parallel cuts taken through a right circular cylinder at an angle other than 90 degrees with respect to the axis of symmetry of said cylinder, each flux plate element having front and back faces that are substantially elliptical, and having major and minor axes. In an embodiment, the flux element angle with respect to the axis of symmetry is substantially 45 degrees. In an embodiment, multiple rotors are attached to a common shaft, or independent shafts coupled through a clutch or similar selectably engageable coupler, and the rotor flux path elements are arranged on said common shaft such that the major axes of the flux path elements are equally spaced on the shaft and wherein the stator poles are in the same position with respect to the common shaft for each motor segment. In another embodiment of this arrangement, the motor has two motor segments and two rotor flux path elements and the rotor flux path elements are arranged on the common shaft such that their major axes are spaced 90 degrees apart.

[0078] In a further embodiment, the motor has rotor counterweights to statically and dynamically balance the mass of the rotor flux elements.

[0079] In a further embodiment, the motor has starter windings adapted to start the motor in a desired rotational direction.

[0080] In a further embodiment, current generated in the windings from collapsing magnetic fields is captured and used.

[0081] One advantage of the presently disclosed system and method is that it addresses the drawbacks of existing systems.

[0082] Another advantage of the presently disclosed system is to provide a direct current motor which develops a significantly reduced Speed Voltage (Vs) component of the Back EMF.

[0083] Another advantage of the presently disclosed system is to provide a direct current motor which makes use of a plurality of salient poles within its stator structure that may possess characteristics different than typically employed by existing Speed Voltage dependent systems. For example, the stator poles should be arranged or constructed to be protected from flux movement in two directions in order to minimize eddy currents, and related iron losses. For example, fabricating all or part of the pole pieces from different metals, using grain orientation, using ferrite materials, using distributed air gap materials, or laminations disposed at right angles with respect to one another, are some techniques that may be implemented to inhibit the production of eddy currents, and thereby lessen iron losses.

[0084] Another advantage of the presently disclosed system is to provide a direct current motor which employs a uniquely shaped rotor having a constant air gap with respect to the salient pole pieces. The constant air gap contributes to a smaller rate of change of inductance in the magnetic circuit, thereby reducing the speed voltage component Vs.

[0085] Another advantage of the presently disclosed system is to provide a direct current motor which employs a shaped rotor having no coils, windings, conductors or bars within its structure. This also contributes to a lower speed voltage component Vs of the Back EMF.

[0086] Another advantage of the presently disclosed system is to provide a direct current motor whose operation is governed by controller, such as an electronic controller, on designed as to orchestrate, synchronize, and control all the internal functions of the direct current motor.

[0087] Another advantage of the presently disclosed system is to provide a direct current motor with a surplus of salient pole windings which are configured to store re-usable magnetic energy within the stator power coil windings. The surplus windings arise from the additional windings possible with the presently-disclosed designs compared to the amount of windings on a similar capacity, traditionally designed motor.

[0088] These and other advantages are achieved in the presently disclosed system by providing a unique arrangement of stator and rotor geometries in conjunction with an electronic controller such that rotation is achieved by means of reluctance switching, synchronized by a position sensor, and acting in response to an electronic controller such that motor input power is properly managed and directed so as to produce a continuous rotation, while simultaneously recovering unused energy momentarily stored within the stator windings.

[0089] One embodiment of the presently disclosed system employs a rotor fabricated from a stack of steel disks, chemically insulated from one another to discourage and reduce eddy currents. The disks may be pressed upon an arbor which, in turn, is obliquely disposed with respect to the intended axis of rotation, and suitably machined on as to produce an assembly with a peripheral contour generally equivalent to that of a cylinder. The stator may be composed of a plurality of salient pole sets, each set comprising a pair of poles, and associated windings, arranged 180 degrees apart from one another upon the stator, and each pole set angularly displaced from one another by a desired number of mechanical degrees.

[0090] In some embodiments, each pole set may also be provided with a concave pole face, whose radius is slightly greater than the radius of the rotor. The rotor, therefore, defines air gap of continuous dimension when rotated. The rotor is in magnetic series with each set of magnetic poles, thereby completing the magnetic circuit, and the rotor reacts to each set of energized poles by undergoing a mechanical displacement equal in degrees to the pole set's mechanical distribution around the periphery of the stator assembly. As the rotor rotates, the zone in which the flux is coupled to the active pole pieces may vary in position along the length of each pole face. However, the width of the air gap separating the pole face from said rotor will not vary.

[0091] This arrangement permits the magnetic potential within the air gap to remain substantially constant, thereby minimizing the change in induction which would normally give rise to the development of a large Speed Voltage (Vs). A greatly reduced Speed Voltage allows a reduced Back EMF in this embodiment of the disclosed direct current motor.

[0092] Other aspects and advantages of the presently disclosed systems and methods will now be discussed with reference to the drawings.

BRIEF DESCRIPTION OF THE DRAWINGS

[0093] FIG. 1 is a plot of flux versus current of air gap energy components in a typical prior art device.

[0094] FIGS. 2A and 2B are cross sectional views illustrating a change in air gap for a prior art device.

[0095] FIG. 3 is a plot of flux versus current for the linear energy relationship in the air gap for the prior art device shown in FIGS. 2A and 2B.

[0096] FIGS. 4A and 4B are equivalent schematic circuits for a prior art DC motor illustrating the steady-state and in-rush operation circuit values.

[0097] FIG. 5 is an overall view of one embodiment of the invention, showing stator sections in cut-away views revealing the disposition of bearings, common output shaft, rotor assemblies, counter weights, stator power windings and stator laminations.

[0098] FIG. 6 is a schematic diagram of an individual rotor/stator section, depicting the relationships between such components as rotor geometry, magnetic flux, air gaps, salient poles and power windings in accordance with some embodiments.

[0099] FIG. 7 is a schematic diagram showing maximum and minimum rotor cross-sections relative to air gaps, stator poles and magnetic circuits in accordance with some embodiments.

[0100] FIG. 8 is a block diagram of an exemplary motor system, depicting forward and rear motor sections, the motor load, the shaft position sensor, the electronic controller and the sump resistor in accordance with some embodiments.

[0101] FIG. 9 is a diagram of a single-rotor with a constant air-gap in accordance with some embodiments.

[0102] FIG. 10 is a diagram of a parallel output cluster of motor sections such as the one shown in FIG. 9 in accordance with some embodiments.

[0103] FIG. 11 is a motor coil energizing scheme for the motors of FIG. 10 in accordance with sonic embodiments.

[0104] FIG. 12 is a schematic of coil interconnections for eight motor sections mechanically connected in parallel in accordance with some embodiments.

[0105] FIG. 13A is a diagram of a motor cluster having brushes and commutator for timing in accordance with some embodiments.

[0106] FIG. 13B is a diagram of a motor cluster having an optical encoder for timing in accordance with some embodiments.

[0107] FIGS. 14A and 14B are schematic cut-away views of a rotor and stator pole pair in accordance with some embodiments of the invention.

[0108] FIG. 15 is an illustration of the non-linear curves representative of the flux behavior as might be measured within a structure of electrical steel of a prior art motor with a variable air gap.

[0109] FIG. 16 is an illustration of the non-linear curves representative of the flux behavior as measured within a structure of electrical steel of the constant air gap motor of the instant disclosure (e.g., FIGS. 14A-14B).

[0110] FIGS. 17A and 17B are schematic representations of a Transformer Voltage (Vt) dependent system in accordance with some embodiments of the present invention.

[0111] FIG. 18 is a schematic illustration of a DC motor system in accordance with some embodiments of the disclosed inventions.

[0112] FIGS. 19A and 19B are schematic illustrations of magnetic flux, electric field, and velocity components within stator iron.

[0113] FIGS. 20A and 20B are schematic end view and side views of certain stator components in accordance with some embodiments of the disclosed inventions.

[0114] FIG. 21 illustrates a conceptual diagram of the generation of an ellipse that, when rotated, has a circular cross-section.

[0115] FIG. 22 is a depiction of some embodiments of the direct current motor shaft assembly.

[0116] FIG. 23 is a cutaway view of some embodiments of a six pole motor stator with associated windings in place.

[0117] FIG. 24 is a cutaway view through the vertical axis of some embodiments of the stator assembly.

[0118] FIG. 25 shows the same cutaway view of some embodiments of the stator assembly shown in FIG. 24, however the rotor has been advanced in angular rotation by 90 mechanical degrees.

[0119] FIG. 26 illustrates a block diagram of some embodiments of an Open Power System Configuration of the direct current motor system.

[0120] FIG. 27 illustrates a block diagram of a Closed Power System Configuration of some embodiments of the direct current motor system.

[0121] FIG. 28 illustrates a logic flow diagram of the functioning of the electronic controller designed to operate with some embodiments the presently disclosed direct current motor. In this case, the logic applies to the operation of one embodiment of an Open Power System Configuration.

[0122] FIG. 29 illustrates a logic flow diagram of the functioning of the electronic controller designed to operate with some embodiments of the presently disclosed direct current motor. In this case, the logic applies to the operation of one embodiment of a Closed Power System Configuration.

DETAILED DESCRIPTION

[0123] In the following detailed description, reference is made to the accompanying drawings that form a part hereof, and in which is shown by way of illustration specific embodiments in which the invention may be practiced. These embodiments are described in sufficient detail to enable those skilled in the art to practice the invention, and it is to be understood that other embodiments may be utilized and that various changes may be made without departing from the spirit and scope of the present invention. The following detailed description is, therefore, not to be taken in a limiting sense.

[0124] FIGS. 5-8 illustrate one embodiment of the motor disclosed herein. Reviewing FIG. 5, it will be seen that the motor consists of a doable stator housing (1, 2) physically separated, but functionally joined together by means of a continuous shaft (10), upon which are mounted two armatures (3, 4), one within each stator assembly. The shaft is carried by bearing sets (11), located within end-bells (14, 15).

[0125] Rotor assemblies (3, 4) each consist of a stack of silicon steel laminations (9), a molded ferrite core, or any other high permeability magnetic material designed to suppress eddy currents, and machined so as to produce a section of a right circular cylinder canted at an angle of 45 degrees with respect to the motor shaft (10). When viewed face on, the rotor structure appears to be elliptical in shape. However, the side view depicts a rhomboid tilted at 45 degrees. This angle may not be the most optimal angle, and it should be realized that other angles may be employed without departure from the spirit of the invention.

[0126] The common shaft (10) may also carry counter weights (7, 8), as depicted, which function to ensure a smooth rotary motion by suppressing mechanical vibrations produced by the uneven mass distribution of the elliptical armature sections (3,4). In another embodiment, each motor segment may include a clutch (25), or some other selectablely engageable coupler in order to couple independent shafts into a common shaft (10). Of course, as many motor segments from one on upwards can be coupled in this, or a similar, manner.

[0127] Each stator assembly contains an individual stack of stator laminations (16, 17) or a magnetic ferrite cylinder, from which extend two or more salient pole projections (12, 13), each of which is wound with a power coil (18). The face of each pole projection (5, 6) is extended to the right and the left of center to ensure continuous air gaps of constant dimension (19, 20), which are aligned parallel to the rotor's edge contour regardless of its angular disposition. Those familiar with the art will realize that it may be possible to install more than two pole projections per armature without departing from the spirit of this invention. Under these conditions, the motor will, of course, operate with a single rotor.

[0128] The pole projections in each stator section are parallel to each other, but the rotor sections are displaced upon the shaft by a predetermined mechanical angle: 90 degrees for two pole sets 120 degrees for three pole sets, etc.

[0129] The motor shaft extends several inches beyond the end bell housings (14, 15) on each side of the motor. One end of the shaft is utilized as a take off point for mechanical power, or load, while the other side of shaft carries a shaft position indicator (21), which is an angular transducer, and may consist of a simple rotary encoder, or a more complex device containing discrete optical sensors and slotted disks.

THE MEANING OF UNITY IN ENERGY CONVERSION SYSTEMS

[0130] The stator power windings may be connected in series or in parallel as preferred. The windings receive their drive pulses from switching transistors, MOSFETs, or other solid state switching devices within the controller (22), which in turn receive their firing instructions directly, or indirectly, from the shaft position sensor (21).

[0131] Power resistor (23) is used as a sump to harmlessly dissipate any remaining energy associated with the collapsing magnetic fields within the stator as the motor rotates.

A Description of the Rotor Geometry

[0132] Drawing attention now, to FIG. 6, it will be noted, that a cylindrical outline is depicted between the poles of an electromagnet, through which the lines of flux are directed in a upward fashion. Notice also, the solid, elliptical lines shown. These demonstrate the shape of the lamination stack or ferrite core which comprises part of this invention. The shape is described by the result of making two parallel slices through a right circular cylinder at an angle of 45 degrees, and then removing all of the cylindrical body except the elliptical core, as demonstrated.

[0133] Magnetically, this elliptical rotor has some very interesting properties. FIG. 7 illustrates a schematic cross-sectional view of the flux path of the rotor in two mechanical positions, each 90 degrees apart. Note, in FIG. 7A, that the elliptical cross-section presents a longer path to the magnetic flux than does the cross-section illustrated in FIG. 7B. Note as well that these figures represent approximate flux paths and not actual cross sectional views of the rotor.

[0134] Accordingly, the elastic nature of the lines of flux will tend to exert a torque upon the rotor geometry, forcing the assembly to rotate 90 degrees, whereby the shortest path is available for the magnetic lines to complete their circuit as is evident in FIG. 7B.

[0135] This process does not require the presence of a "secondary" magnetic coil, the addition of which would tend to decrease a motor's overall inductance, by means of quadrature coupling, or armature reaction, during normal operation.

DETAILED DESCRIPTION OF THE MOTOR'S OPERATION

[0136] One embodiment of this invention employs two rotors, each fabricated from a stack of laminated disks, pressed upon arbors which are obliquely disposed with respect to the intended axis of rotation, and then integrally machined in order to provide both rotors with peripheral contours equivalent to that of a cylinder while retaining their overall elliptical shape. Each stator section is formed by a lamination stack having two, spaced-apart, salient pole projections terminating in concave pole faces whose radii are slightly larger than the radius of each rotor. Both rotors thereby define air gaps of constant dimension while rotating. Each rotor is in magnetic series with two air gaps and two pole pieces and a complete magnetic circuit which contains its own coils for the production of magnetic flux. Each magnetic rotor circuit is separate and distinct from each other magnetic rotor circuit, although they share a common output shaft. An angular position sensor or shaft encoder is positioned at one end of the output shaft, and sends electronic position signals to a DC power supply/controller, which in turn sends pulses to the motor stator sections as required.

[0137] The application of a current pulse to a given set of stator coils causes the rapid rise of magnetic flux within the selected stator section and its associated rotor. The increased flux density then causes the rotation of the active rotor, as the flux lines "shrink" to ensure their manifestation in a circuit of minimum length. The output torque is produced by the laws of magnetic reluctance acting in conjunction with the innovative geometry of the rotor. No current carrying conductors are involved in the rotor.

[0138] As the first rotor reaches its position of minimum cross-sectional diameter, the shaft encoder then directs the electronic controller to send a power pulse to the second rotor, and the operation repeats itself. When this procedure is enacted every 90 degrees, the result is a smooth angular rotation, and the production of a continuous average torque. However, a secondary result of this arrangement is the production of an electrical output from each stator section as a result of the collapsing of its magnetic field at the end of each power cycle. This electrical energy may be harmlessly dissipated in a sump resistor, or it may be put to use, for example in powering other devices, including lamps or heaters or recovered to supply a portion of the energy used to drive the motor.

[0139] In an embodiment, an exemplary motor utilizes a rotor geometry consisting of a lamination stack or a molded ferrite shape, canted at a specific angle with respect to the output shaft, while retaining a circular cross section to the axis of rotation, and presenting an overall elliptical appearance in its own plane. This arrangement allows for a constant air gap to be maintained between the rotor's edge and the pole pieces thereby producing mechanical torque without the utilization of coils or conductors residing anywhere upon said rotor.

[0140] One embodiment of the motor employs a plurality of "elliptical" rotors mounted upon the same output shaft, but positioned such that each rotor section is advanced a certain number of mechanical degrees from the others such that torque production over 360 degrees of rotation is shared equally by the number of rotors utilized. The motor also has a plurality of pole sets and separate magnetic circuits, such that each elliptical rotor section is associated with its own external source of magnetic flux, regardless of the fact that they share a common output shaft. Accordingly, the salient stator pole projections will all reside in the same plane and be parallel to each other, while the rotor sections will be displaced upon the output shaft by predetermined mechanical angles; 90 degrees for two pole sets, 120 degrees for three pole sets, etc. Those skilled in the art will realize that this arrangement may be reversed without departing from the spirit of the invention. Likewise, those skilled in the art will also realize that it is possible to construct a single, standalone, motor utilizing a single rotor and stator section.

[0141] Referring now to FIGS. 5 and 7, which each depict the relationship of the rotors to the stators, it will be noted, that the left hand rotor is positioned between the salient poles of its stator such that its oblique length presents the longest possible path to the magnetic flux produced by the associated pole set. The right hand rotor on the same shaft, will simultaneously present its shortest cross sectional path to its associated pole projections.

[0142] Sensing this arrangement, the shaft position sensor (21) will cause the controller (22) to energize starting windings (not shown) which will rotate the motor shaft in the desired direction, while simultaneously sending a current pulse into the left hand pole set depicted in FIG. 5. Those skilled in the art will understand and appreciate how starter windings are implemented to start a motor in the desired rotational direction.

[0143] The appearance of lines of force within the first rotor segment will cause a twisting action upon that rotor's lamination stack, such that torque is produced upon the motor output shaft in the desired direction. At the same time, the right hand rotor is rotated, by the turning shaft, into a position of readiness with respect to the right hand magnetic pole set.

[0144] The shaft position sensor (21), illustrated in FIG. 8, then signals the controller (22), which directs a current pulse into the second stator pole set, advancing the output shaft by another 90 degrees. Utilizing this means, each motor half is alternately energized and a complete revolution of the shaft is achieved with every four electrical pulses Thus a 900 RPM motor will require: 4 Pulses/Rev×900 Rev/Min.=3600 Pulses/Min supplied from the controller's power supply.

[0145] The average torque available on the motor output shaft will be a function of the cooperative effort developed by both rotors over each mechanical revolution. The output torque developed by this method is strictly a reluctance torque, generated as the lines of magnetic flux within each rotor section alternately shrink in an attempt to provide themselves with the shortest possible magnetic path between poles.

[0146] It is important to realize that this torque-producing mechanism does not involve any interaction of either stator's magnetic field with a current carrying conductor of any kind, neither in the form of a Speed Voltage interaction, nor in the form of a transformer coupling with a time-varying field. Instead, the torque appearing on the motor shaft is a direct function of the rotor's geometry interacting with forces produced at the boundaries between the rotor body and the stator poles, and by internal cam action particular to the rotor geometry in the presence of a contracting flux.

[0147] Magnetic energy stored in the stretched lines of flux between each pole set must be dissipated as each field structure collapses in response to instructions from the controller. This will ensure that an "empty" inductor wilt be available at the start of each 90 degree cycle. Accordingly, fly-back diodes are provided in association with each power winding. The diodes direct pulses generated by the collapsing fields into a sump or load resistor (23), where they may be harmlessly dissipated as excess heat. Alternatively, said energy may be used to power other electrical appliances external to the motor, or may be applied to a capacitive storage element and then utilized to send power back to the main power supply.

Efficiency and Scaling

[0148] Because of the rotor geometry, in conjunction with the fact that this type of reluctance motor carries no rotor windings, at least 50% of the I squared R losses, stray copper losses and hysteresis losses experienced by traditional motor technology will be avoided in accordance with the spirit of invention.

[0149] Energy savings of this magnitude are possible primarily because of the constant air gap afforded by the rotor's geometry. However, it should be remembered, that any electromagnetic device so designed as to prevent a large change in the reluctance of its magnetic circuit, while ensuring a constant air gap during the course of any mechanically sponsored alteration in the mean circuit length, shall experience only minute variations in inductance. The operational benefits of such an arrangement will be that any force produced or work done by the electro-mechanical process, will have a minimal effect upon the magnetic excitation current.

[0150] Additionally, the use of high frequency switching technology to develop the required pulses of drive current will ensure that conversion efficiency, or the transformation from electrical power to mechanical power, will be attainable in the high 90 percentile range.

[0151] Application of concepts herein disclosed may be arranged such that the rotor segments may be joined either in series, as depicted in FIG. 5, or in parallel, such that each rotor is equipped with a gear upon its output shaft, and several such assemblies are situated so as to drive a common gear and a main output shaft, or with single rotors in multi-pole embodiments. This adaptability is possible in series and parallel arrangements.

[0152] The scaling of these embodiments is relatively straightforward. Accordingly, no unusual difficulties are anticipated in producing small, medium or very large sized motors of this design.

[0153] in another embodiment, an electric motor cluster comprises several stator sections each possessing a minimum of two salient pole projections, wound with power windings, and each having a single armature rotor. Each individual rotor is angularly displaced one from the other, while mounted upon a common frame, and geared together such that each motor section contributes to the rotation of a common output shaft. Those skilled in the art will also recognize that it is possible to deploy a single, standalone motor with a single rotor and stator pair rather than as part of a cluster.

[0154] Such an arrangement not only allows for the combining of motor output powers and the removal of flutter from the final mechanical output, but simultaneously allows for a large increase in output torque by virtue of the necessary reduction gearing. The embodiment suggested within this particular disclosure lends itself perfectly to applications within the field of electric vehicle propulsion, particularly in those cases where the prime mover is to be located within the wheels of the vehicle. However, other applications are easily envisioned.

[0155] Each motor section shall consist of stator and armature elements as described in PCT application number PCT/US09/46246, filed on Jun. 4, 2009, and entitled "PULSED MULTI-ROTOR CONSTANT AIR GAP RELUCTANCE MOTOR." The motor may consist of the following features:

[0156] A stator, consisting of a stack of laminations, or a molded ferrite core, so constructed as to provide at least one set of salient magnetic poles, spaced apart 180 mechanical degrees, and situated so as to allow an air gap to exist between the stator structure and the armature of the motor. Each salient magnetic pole projection may be wound with power windings, the function of which is to produce a magnetic field of considerable strength, and direct the same through the air gaps and into the body of the motor's armature.

[0157] An armature, also consisting of a stack of laminations, or a molded ferrite shape, so designed as to present each set of field poles with a cylindrical contour, perceived beyond each air gap, while retaining an elliptical profile with respect to the output shaft. The armature sections carry no electrical windings of any kind, and require no slip rings or, field coils or permanent magnets. However, armature segments may require shaft-mounted counter weights to offset their eccentricity, and maintain angular balance during rotation.

[0158] The power windings wound upon the salient pole projections, are energized by pulses of electric current produced by a DC power supply and provided through an electronic controller unit, or through a mechanical commutator, etc. The pulses are automatically applied to the salient pole nearest the longest flux path available through a particular rotor section, as determined by a shaft position sensor, or the geometry of a commutator.

[0159] The appearance of flux lines linking any stator pole set and any armature section immediately causes a rotation of the motor's output shaft by 90 mechanical degrees as the flux lines seek to establish the shortest possible path available for the completion of their magnetic circuit within a given motor.

[0160] This action is transmitted to the main output shaft via a large reduction gear, thereby increasing the available torque. In the motor cluster embodiment disclosed herein, several motor sections are positioned such that each may contribute to a common mechanical output. However, several such motor sections may be energized simultaneously, thereby increasing the output power in multiples.

[0161] Upon detecting motion, the shaft position sensor communicates the change in position of the output shaft to the electronic controller, and current flow is then terminated in each active stator section, and instantly initiated in the stator section windings next scheduled to be activated. By means of such switching action, which occurs at even intervals of mechanical degrees, a constant rotary motion is ensured.

[0162] FIGS. 9-13 illustrate one embodiment of the motor cluster disclosed herein. Reviewing FIG. 9, it may be seen, that each motor section consists of a metallic housing 1 containing a stator stack 16 and an armature assembly 3, which is mounted upon an output shaft 10, which is carried by two sets of bearings 11, located within end bells 14.

[0163] The rotor assembly 3 within each motor section, consists of a stack of silicon steel laminations 9, or a molded ferrite of appropriate shape, or any other high permeability magnetic material designed to suppress eddy currents, machined so as to produce a section of a right circular cylinder canted at an angle of 45 degrees with respect to the motor output shaft 10. When viewed face on, the rotor structure appears to be circular in shape. However, the side view depicts an ellipse tilted at 45 degrees. This angle may not be the most optimal angle, and it should be realized that other angles may be employed without departing from the spirit of the invention.

[0164] Each motor shaft 10 may also carry counter weights 7, as depicted, which function to ensure a smooth rotary motion by suppressing mechanical vibrations produced by the mass distribution of the eccentric armature design 3. Each motor shaft carries a high speed output pinion 24 which is designed to mesh with the main output gear as shown in FIGS. 9 and 10.

[0165] Each stator assembly contains an individual stack of stator laminations 16 or a magnetic ferrite cylinder, from which extend two or more salient pole projections 12, each of which is wound with a power coil 18. The face of each pole projection 5 is extended to the right and the left of center to ensure continuous air gaps 19 of constant dimension. The pole faces are aligned parallel to the rotor's edge contour regardless of its angular disposition. Those familiar with the art will realize that it may be possible to install more than two pole projections in association with each armature without departing from the spirit of this invention.

[0166] Referring now to FIG. 10, the concept of the parallel motor cluster will become apparent in greater detail. The embodiment depicted makes use of eight individual motor elements numbered clockwise, M1 through M8, starting at the 9:00 o'clock position. The motor elements are mounted at 45 degree intervals upon a circular frame 61. Each motor element consists of a laminated, four pole stator stack 62, an air gap 68, an elliptical rotor 67, an individual motor output shaft 64, and an output pinion 63. Further, it will be noted, that each output pinion is in mesh with a central output gear or "bull gear" 65 which drives the main output shaft 66.

[0167] This arrangement allows for four motors to be energized at any one time, with power overlaps and torque-sharing occurring at 45 degree intervals. This feature serves to smooth out the total torque delivered to the output shaft, allowing for a more continuous delivery of power, as each contributing motor develops its output torque out of phase with respect to each of the others. Total motor action during operation may be appreciated by studying the coil energizing truth table depicted in FIG. 11, while the power coil interconnection schematic may be reviewed in FIG. 12. In FIG. 11, the horizontal portions of each chart depict energized coils and the sloped portions of the chart represent the magnetic reset of the energized coils. There are shown coil sets for eight motors as described in the above text with respect to FIG. 10.

[0168] Referring now, to FIG. 12, it will be noted that switches S1A through S8A, and switches S1B through S8B, are used to control the power winding coil sets in each motor section. The coil sets are labeled A, A' and B, B' for each motor as shown in FIG. 10. These switches are schematically accurate, but may represent either solid state switching devices located within the electronic motor controller, or actual contact bars located upon a more traditional commutating device. These distinctions are more clearly explained in FIG. 13.

[0169] FIGS. 13A and 13B depict two variations of some embodiments of the present invention. FIG. 13A demonstrates the parallel motor cluster concept employing a traditional electro-mechanical commutating device 56, 57, while FIG. 13B demonstrates a more modem approach employing a shaft-mounted encoder 59, a micro-processor, and an electronic motor controller. It will be noted, that both systems require a source of DC power, as well as a capacitive power sump 58, into which excess "inductive energy" is directed. This "sump" may be equipped with a resistive load, which will consume said inductive energy, or the accumulated potential may be utilized to supply other worthwhile power requirements.

[0170] Returning now to FIGS. 13A and 13B, it will be noticed that each arrangement contains a motor cluster housing 51, a plurality of high speed motor pinions 52 mounted upon individual motor output shafts 53, and a central bull gear 54 mounted upon a main output shaft 55. However, FIG. 13A makes use of a mechanical commutation device 56 with standard carbon brush contactors 57, while the device shown in FIG. 13B employs a shaft encoder 59 and an encoder pick-up device 60.

[0171] Observing FIG. 13B, it will be noted that electronic signals obtained from the encoder assembly are transmitted to the micro-processor and the electronic motor controller, while power pulses are independently directed to individual motor windings via output conductors energized by the motor controller. Alternatively, the arrangement shown in FIG. 13A accomplishes these functions electro-mechanically, which may be advantageous in situations requiring the control of electric power greater than can be managed by present day solid state switching devices. Ultimately, however, both sys-

THE MEANING OF UNITY IN ENERGY CONVERSION SYSTEMS

tems produce the results depicted in FIG. **11**, and both systems ultimately direct inductive energies from collapsing magnetic fields into the capacitive sump indicated by network **58**.

[0172] It should be understood that the embodiment discussed in this application and depicted in associated FIGS. **9-13**, are for illustrative purposes only, and that those having skill in the electrical arts will understand that modifications and alterations can be made hereto, within the spirit of the present invention.

[0173] As discussed previously, the parasitic effect of Back EMF, and motors designed to exploit Speed Voltage (Vs), imparts several drawbacks to existing systems. At least in part to avoid these and other drawbacks, the presently disclosed systems and methods are designed to operate on the production of Transformer Voltage (Vt). As disclosed herein, at least one advantage of such a design is that it allows the energy associated with the magnetic field to be re-captured and, in great measure, re-utilized.

[0174] To exploit the Transformer Voltage (Vt) instead of the Speed Voltage (Vs), the presently disclosed systems and methods implement the Wowing two design principles arising out of the above discussion, and an understanding of the importance of equation 6 above. The first design principle implemented to exploit Transformer Voltage (Vt) is to introduce a parameter dl/dt corresponding to the change in magnetic circuit length over time. The second design principle is that to minimize the Speed Voltage (Vs) component the relation provided in equation 8 must be zero, or nearly zero. One way to accomplish a nearly zero Speed Voltage (Vs) is to minimize dL/dt by designing the air gap to be constant. These two design principles are described in greater detail below.

[0175] The consideration of the change in magnetic circuit length over time (dl/dt) can be described with reference to FIGS. **14A** and **14B** which are schematic cut-away views of a rotor and stator pole pair in accordance with some embodiments of the invention. As shown in FIG. **14A**, stator poles **500** form a pair on either side of rotor shaft **510**. Magnetically conductive rotor stack **520** is mounted on shaft **510** and depicted in a first position in FIG. **14A**. In the embodiment depicted, rotor stack **520** may comprise a shape that is designed to present a substantially cylindrical profile when rotated about shaft **510**. For example, and as described in more detail below, rotor stack **520** may comprise a substantially elliptical shape that is mounted on shaft **510** in an offset, or canted, fashion forming an angle θ with respect to the shaft **510** as best seen in FIG. **14A**. As also depicted, in the position shown in FIG. **14A**, rotor stack **520** forms an air gap of distance g1 with stator poles **500**. The magnetic circuit formed by the stator poles **500** and rotor stack **520** can be calculated from adding the air gap to the major-axis length l_1 of the rotor stack **520** as follows:

[0176] FIG. **14A** magnetic circuit length=g1+l_1+g1=2g1+l_1.

[0177] FIG. **14B** shows a cross sectional view when the rotor stack **520** is rotated one-quarter turn (i.e., 90 degrees) from the position shown in FIG. **14A**. As shown by comparison with FIG. **14A**, and by design, the air gap in the FIG. **149** position (g2) between rotor stack **520** and stator poles **500** remains constant (i.e., g1=g2), however the length of the magnetic circuit in FIG. **14B** is now a factor of the rotor stack **520** minor-axis and can be calculated as:

FIG. **14B** magnetic circuit length=g2+l_2+g2=2g1+l_2.

[0178] Therefore, by design, when the shaft **510** rotates, the magnetic circuit length will vary in time between a maximum proportional to e and a minimum proportional to l_2. Furthermore, as the dimension of the air gap does not change (i.e., g1=g2), the contribution of dL/dt is zero, and the Speed Voltage component is, by design, zero as well.

[0179] The following is a closer examination of the effect of the new parameter, dl/dt, or a change in magnetic circuit length with respect to a change in time in accordance with the disclosed inventions. Beginning from the classical formula for inductance (equation 13):

$$L = (N^2 \mu A)/(Kl),$$

[0180] where N is the number of turns, μ the permeability, A the cross-sectional area, l the magnetic circuit length, and a K constant of proportionality. In most inductance calculations, all of the above parameters are usually considered to be constants. However, as explained above, in the presently disclosed embodiments the length of a magnetic circuit changes in time. Accordingly, it is interesting to examine the magnitude of the resulting change in inductance using the following values determined experimentally by the above-named inventor.

[0181] In one embodiment, measuring a mean magnetic path around the stator equivalent to the mean circumference Cm, gives 43.982 inches in length. A major axis for a rotor stack **520** of 14 inches long gives the total circuit length l_1=57.982 inches. As discussed in connection with FIG. **14B**, rotating the rotor stack **520** by 90 degrees, changes to the minor axis of the rotor stack **520** and also provides an overall circuit length l_2=55.486 inches. Substituting and calculating corresponding values of inductance using equation 13 above gives:

$$L_1 = 0.103480 \text{ Henrys, and } L_2 = 0.104346 \text{ Henrys.}$$

[0182] The difference of these two values ΔL is calculated to be 8.666×10^{-4} H, and when this change occurs in one quarter of a rotation at 60 HZ, a measured Back EMF of 2.5 Volts results. This is a remarkable result, considering the fact that a change of the same degree within the air gap of a conventional Speed Voltage based motor generates hundreds of volts.

[0183] To illustrate the significance of the above result, we compare FIG. **15** and FIG. **16**. In the manner of linear energy for an air gap shown in FIG. **3**, FIG. **15** shows the non-linear curves representative of the flux behavior as might be measured within a structure of electrical steel of a prior art motor with a variable air gap. As shown, plots **600** and **602** are continuous, but quite non-linear. This is to be expected, because here, as in the case of B/H curves, the permeability (μ) is not constant.

[0184] In correlation with the FIG. **3** air gap example, the following calculations illustrate the changes observed in this steel sample as the associated air gap changes from its g1 dimension to its g2 dimension. Again, starting from equation 11,

$$E_T = Id\Phi + \Phi dL$$

[0185] For the values shown on FIG. **15**, for a gap size of g1: I=12 amps, d(I_1)=3.202, and dI_1=8.23. Therefore, E_{T1}=(12)(3.202)+(8)(8.23)=104.26 Joules. For a gap size of g2: I=7 amps, dΦ$_2$=3.475, and dI_2=5.4535. Therefore, E_{T2}=(7)(3.475)+(9)(5.4535)=73.406 Joules.

[0186] Unlike the air gap calculation corresponding to FIG. **3**, here each energy component is different in value, as might be expected. However, note that the total energies, E_{T1} and E_{T2}, are not equal in this case. There is a substantial difference of 30.86 Joules.

[0187] The contrasting, and unexpected result of the present invention is shown in FIG. 16, which is an illustration of the non-linear curves representative of the flux behavior as measured within a structure of electrical steel of the constant air gap motor of the instant disclosure (e.g., FIGS. 14A-14B). Calculating again using equation 11, for the rotor stack 520 in the first position (FIG. 14A): $Id\Phi_1 = (12)(3.475) = 41.70$ joules, and $\Phi dI_1 = (9)(5.4535) = 49.08$, and $E_{T1} = 90.78$ Joules as shown by plot 700. For the second position (FIG. 14B): $Id\Phi_2 = (11.98)(3.475) = 41.63$ Joules, and $\Phi dI_2 = (8.85)(54535) = 48.26$ Joules, and $E_{T2} = 89.89$ Joules as shown by plot 702. Accordingly, the difference in energies is 0.89 Joules.

[0188] As demonstrated above, the difference in behavior here is very distinct from conventional systems: a small decrease in current (I), and an equally small increase in flux (Φ). This can only be possible without the presence of a speed related Back EMF. Accordingly, it stands to reason that the energy usually associated with the Speed Voltage (Vs) has been reduced to a value that cannot possibly support the measured shaft horsepower. However, because the primary relationship for energy in this system is:

$$Mv = Es + Ec,$$

[0189] it also stands to reason that if the co-Energy factor is reduced, and Field Energy remains constant, then there must have been a change in the supply energy. This can be understood by looking at the power involved, rather than from the energy domain. Recalling that the total applied voltage is the sum of the voltage drops around the equivalent motor circuit, we can write:

$$d\Phi/dt = LdI/dt + I\, dL/dt.$$

[0190] where $d\Phi/dt$ is the source voltage, LdI/dt is the Transformer Voltage (Vt), and $I\, dL/dt$ is the Speed Voltage (Vs). Substituting V for $d\Phi/dt$, we obtain:

$$V = LdI/dt + I\, dL/dt.$$

[0191] However, the actual source voltage is the sum of Vt, Vs and Vr, so we must modify the above expression accordingly, thus obtaining:

$$(Vt + Vs + Vr) = (LdI/dt) + (I\, dL/dt) + Vr.$$

[0192] Because Watts are the product of Volts and Amps, the above expression is now multiplied by I to get:

$$(Vt + Vs + Vr)I = I(LdI/dt) + I(I\, dL/dt) + IVr.$$

[0193] From Ohm's law we know that Vr is actually equal to Ir, thus we may substitute:

$$(Vt + Vs + Vr)I = I(LdI/dt) + I(I\, dL/dt) + I^2 r.$$

[0194] Thus, we finally arrive at an expression in Watts which represents the motor in question.

[0195] Recalling the fundamental nature of equations it is obvious that whatever we change on one side of the equal sign, we must change on the other side to maintain a mathematical balance. Accordingly, Vr I must equal $I^2 r$ as the motor losses are constant. If the Speed Voltage parameter I(IdL/dt) is reduced almost to zero, because of rotor geometry and a constant air gap, then it stands to reason that its supporting component (Vs) in the source voltage must also be reduced by the same proportion. This must be so if power, and its associated energy, are to be conserved. Accordingly, it now becomes apparent that the Back EMF is a parasitic agent, the presence of which demands a higher source voltage to perform the same work; Back EMF is a system loss. However, this kind of loss only destroys potential, it does not evolve heat, therefore, it has gone unnoticed until now.

[0196] Another unexpected consequence of the presently disclosed technology resides in the fact that the reluctance torque is not affected. The torque generating mechanism does not care if it is supported by the field energy or the co-energy, it simply responds to the presence of flux according to the formula:

$$T = -\tfrac{1}{2}\Phi^2\, dR/d\theta.$$

[0197] As noted above with reference to FIGS. 4A and 4B, Back EMF causes significant issues in the operational characteristics of a conventional motor. However, under the above-described and currently disclosed embodiments, Back EMI' does not appear in the traditionally anticipated magnitude, but the motor still undergoes an acceleration, Thus, for exemplary purposes, using the values in FIG. 4A and calculating the characteristics while ignoring Back EMF the motor would develop 18.221 HP, or 13,592.866 shaft watts, and would require a total input power of 16,859.670 watts. Subtracting the shall watts from the total input power, the figure of 3,266.804 watts is obtained. Dividing this number by the operating current of 135.965 amps, a potential of 24.02 volts is indicated. However, there is no place for such a voltage in the equivalent circuit diagram used to obtain this information; an indication that something is out of balance in the overall energy distribution. Speed Voltage cannot be missing, because it was stipulated at the start of these calculations that it did not exist. However, one candidate still remains, LdI/dt, or the Transformer Voltage. Checking this assumption is quite a straight forward matter. Using the relationship: V=ldI/dt, assuming an acceleration time of 6/10 seconds, and solving for L, a value of 0.1059 H is derived, which is very much in keeping with the inductance figures described above in connection with FIGS. 4A-4B. Therefore, Vs is not required to power the presently disclosed kind of motor, instead Vt is the driving agent.

[0198] The differences between Speed Voltage (Vs) dependent systems and Transformer Voltage (Vt) dependent systems are many and pronounced. The most pronounced difference between Vt and Vs lies in the inductive mechanism with which each potential is associated. Regarding the term IdL/dt, under dimensional analysis yields that dL/dt has the dimension Joule-seconds/coul^2, which is representative of a resistance. Hence, I^2 (dL/dt) is dissipative by its very nature, while the expression VI, from which LdI/dt is derived, can easily describe a reactive condition. Energy can be extracted from a reactive situation, but not from a dissipative relationship.

[0199] FIG. 17A is a schematic representation of a Transformer Voltage (Vt) dependent system in accordance with some embodiments of the present invention. As depicted, a DC motor 800 has a through-put efficiency of 79.84%, such that a power input 802 of 3,264.424 watts, minus system losses 804 of 658.128 watts, yields an output 806 on the shaft of 2,606.296 watts, or approximately 3.5 HP. Over and above this shaft output 806, the motor 800 supplies an electrical output 808 due to the re-capture ability associated with IVt. Assuming a theoretical 100% recapture is possible, then this output electrical power 808 has a maximum value of 2,606.296 watts. However, in practice, no process can be 100% efficient, and so, a more physically reasonable arrangement is

THE MEANING OF UNITY IN ENERGY CONVERSION SYSTEMS

displayed FIG. 17B where a recapture electrical output **810** figure of 90% is used. As shown in FIG. 17B, the power through-put from the electrical input **802** to the mechanical output **806** remains the same at 79.84%. However, the reclaimed "field energy" now delivers a useful electrical output **810** of 2,345.666 watts.

[0200] The recaptured electrical output **810** power is the same power that was applied earlier (e.g., **802**), minus all the associated losses. In operation, the input power **802** pulse, and the recapture power **810** pulse cannot exist at the same time. They are 180 electrical degrees out of phase with each other.

[0201] FIG. **18** is a schematic illustration of a DC motor system in accordance with some embodiments of the disclosed inventions. As shown, the drive section of the electronic controller **900**, in these embodiments, contains four field poles, and so the controller **900** issues four sequential pulses into the motor every 90 degrees, each pulse containing 816.106 watts. If it is desirable to measure, or otherwise monitor, these pulses, a meter **902** can be implemented as illustrated. In response to the input pulses from controller **900**, motor **904** responds by rotating, and loses 658.128 watts in heat losses **906**.

[0202] The output power **908** available at the motor **904** shaft, may be approximately 3.5 HP, and the overall motor efficiency may be 79.84%, as measured by contrasting total electrical input from controller **900** to the average mechanical output **908** at the shaft.

[0203] Almost simultaneously, each collapsing motor field produces an electrical output **909** of 586.416 watts, which represents the re-captured field energy. These pulses **909** are then delivered to the recovery section **910** of an electronic controller, and then may be stored, for example, in the re-capture capacitor bank **912**. In some embodiments, energy from this capacitor bank **912** could be removed if necessary, and used to supply power to external appliances (shown in phantom at **914**, **916**).

[0204] As power pulses are delivered to the recapture capacitor bank **912**, voltage across these capacitors will begin to rise. Once the potential reaches a certain pre-determined value, the feedback controller **918** may automatically start sending power back to the main capacitor bank **920**. In some embodiments, the power delivered by this motor **904** operation may be monitored by the feedback watt meter **922**.

[0205] A power accounting at this point demonstrates the subtle energy workings at play within this motor system:

(3,264.424 watts)−(Motor Losses=658.128 watts)=
(shaft power of 2,606,296 watts);

[0206] (Recaptured power is 0.9×2,606.296)=(2,345.666 watts, sent to feed-back).

[0207] However, (3,264.424 watts)−(2,345.666 watts)=(918.758 watts), which represents a power shortage. Therefore, this amount must be drawn from an external power source, such as the utility line or source voltage **924**. Because of the unique features of the disclosed embodiments, the system of FIG. **9** also yields the following efficiencies:

[0208] 1.) Overall Motor Efficiency=79.84%; and
[0209] 2.) Apparent System Efficiency=2,606.296 watts/ 918.758 watts×100%=283.676%.

[0210] While this apparent system efficiency is remarkable, it is understandable in view of the above explanation of Transformer Voltage (Vt) operation (and resultant lack of Back EMF). Furthermore, the system inputs and losses are as expected:

[0211] Motor Losses=658.128 watts;
[0212] Recapture Losses=260.630 watts; and
[0213] Total from Line=918.758 watts.
[0214] Thus, the line only supports the system losses, while the shaft power is supported by the change in field energy per unit time. As expected, the motor will not operate without line power.

[0215] As noted herein, the unique characteristics demonstrated by the disclosed DC motor, are the result of a special cooperation between the rotor design and the stator design. With respect to the stator, several design features are important. Therefore, the DC motor as disclosed herein may include combinations of the following features: an even number of salient stator poles, salient poles that are protected from flux movement in two directions, poles that are designed to be as short as possible, and pole windings should be of adequate wire size, but with as many turns as desirable.

[0216] Some reasons and advantages of the above-noted stator design features are the following. The even number of salient poles is advantageous in establishing the flux field to impart a force on the rotor, because each pole set constitutes a complete magnetic circuit for each phase with two poles being the minimum set.

[0217] As explained herein, and with reference to FIGS. 19A and 19B, the disclosed motor will experience two flux movements within the motor. FIG. 19A is an illustration of a portion of some stator lamination plates **1010** in accordance with some embodiments of the disclosed motor. Each lamination plate **1010** may also comprise an insulating coating **1012** on the outer surfaces. As shown, a magnetic flux field **1014**, indicated as coming out of the page by the dots as shown, experiences a first velocity (v_1) indicated by arrows **1016** pointing to the right, and an electric field (E_1), indicated by the arrows **1018** pointing to the top of the figure. This field (E_1) produces a relatively insignificant eddy current because the insulating coating **1012** between each plate inhibits the current flow. However, as shown in FIG. 19B, when a second direction of motion (v_2) is experienced as indicated by the arrows **1020**, such motion will produce a second electric field (E_2) as indicated by the arrows **1022**. Because this field (E_2) is established between the insulating coatings **1012**, eddy currents (I) as indicated by arrows **1024** will flow within the metal lamination plates **1010**.

[0218] FIGS. 20A and 20B illustrate an end view and a side view of stator pole arrangements in accordance with some embodiments of the disclosed motor that enable the minimizing of the eddy currents in the salient poles due to flux movement in two directions as described above. As shown for this embodiment, a stator pole may comprise a top pole piece (called a shoe) comprising vertically disposed laminations **1028**. A bottom portion of the pole may comprise standard, or radially disposed, laminations **1030**. Other arrangements of laminations are also possible, the concept being that the layers of the various portions are arranged to minimize eddy currents by inhibiting current flow.

[0219] Also illustrated for this embodiment in FIGS. 20A and 20B are stator windings **1026** for generating the magnetic flux fields, rotor **1032**, rotating about an axis of rotation **1034**, and constant air gap **1036** between the edge of rotor **1032** and stator shoe **1028**.

[0220] Additional embodiments of stator poles may also be implemented to minimize eddy currents. For example, another embodiment is to have the pole face, or shoe **1028**, made of a material such as sintered steel, ferrite, or distributed air-gap material, and then bond, or otherwise fasten, the shoe **1028** to the bottom portion **1030** of the stator pole. Likewise, other embodiments may also implement stator pole pieces comprising grain-oriented steel, with the grain best oriented for lateral flux movement. Embodiments employing combinations of these techniques for eddy current minimization are also possible.

[0221] Likewise, for some embodiments, the salient poles are designed to be as short as is optimal in order to optimize the overall magnetic circuit length. This has the advantage of also lessening motor iron losses.

[0222] Finally, for some embodiments, the design of the pole windings (e.g., windings **1026**) is to be of adequate wire size, but with a number of turns that is optimal. This has the advantage of keeping I^2R (i.e., copper) losses to a minimum. The wire size and number of turns are preferably optimized so that enough turns are used to establish a magnetic flux of sufficient magnitude, while also keeping the I^2R losses to an optimal minimum. Typically, relative to a comparable Speed Voltage dependent motor, the presently disclosed stator designs will accommodate a greater number of windings per pole.

[0223] As noted previously, the rotor design features of the presently disclosed invention also contribute to the herein described performance. As discussed above, an important feature of the disclosed rotor is that it be shaped to assist in the reduction of the factors that contribute to the generation of Back EMF. To that end, rotors that exploit Transformer Voltage (Vt) in accordance with the present disclosure will be designed to form a constant, or substantially constant, air gap with respect to the stator poles.

[0224] In addition, a rotor designed to exploit Transformer Voltage (Vt) in accordance with the disclosed embodiments of the invention will also facilitate the creation of a variable length magnetic circuit path. In general, one way to design a rotor capable of creating a variable length magnetic circuit path is to create an ellipse that, when rotated, has a circular cross-section. For some embodiments, such an ellipse may be created in the manner illustrated in FIG. **21**.

[0225] FIG. **21** illustrates a conceptual diagram of the generation of an ellipse that, when rotated, has a circular cross-section. Such an ellipse **1000** can be generated by drawing a reference circle c with a radius r. Projecting out of the plane of the circle c, a height h is generated from $r \sin \alpha$, where α is that angle of inclination of the hypotenuse R (of triangle a0b) from the plane of circle c, and where θ represents the angles generated about the point 0 in the plane of circle c. Thus, the triangle a0b is formed having a value of $R=(r^2+(r\sin\alpha)^2)^{1/2}$. Further, $R=r(\cos \alpha)^{-1}$. If the height (h) of the triangle a0b is varied sinusoidally in accordance with the angle θ, then for a given θ, $R=r (\cos \alpha)^{-1} \sin \theta$. Plotting an infinite number of similar triangles about θ for the full 360 degrees of circle c produces an ellipse e_p of perimeter e_p, as shown in FIG. **12**. Ellipse e_p will always have a circular cross-section when rotated about 0 in the plane of circle c. Additional rotor designs suitable for implementation of the concepts presently disclosed are also possible.

[0226] Having described the relevant design features for the stator and the rotor, we turn now to a description of some embodiments of the instant DC motor system. Traditionally, a DC motor consists of three main components: a stator assembly for supporting the magnetic field coils, a shaft-mounted armature, or rotor, for supporting windings of its own, and a commutator, also shaft-mounted, which supplies a timed switching function by means of two or more carbon brushes for controlling the supply electrical current to the rotating armature assembly from an external power supply.

[0227] FIGS. **22** through **25** show aspects of some embodiments of the presently disclosed DC motor. FIG. **22** illustrates one embodiment of the motor's rotor assembly **1190**, wherein **1100** is the shaft, **1101** are bearings, **1102** depicts rotational stabilizers, or counterweights, desirable to offset any eccentricity of the magnetically conductive lamination stack **1104**, which may be mounted upon an arbor **1103**. The rotor assembly **1190** may also contain a shaft position sensor **1108**, which may consist of a mounting hub **1105**, and one or more encoded disks **1106**. Positional information carried by the disks, is read by sensor heads **1107**, and an appropriate signal is conveyed to the electronic controller **1503** (shown in FIG. **26**, but not shown in FIG. **22**), for interpretation and generation of electronic control commands. Other embodiments of the rotor assembly **1190**, the shaft position sensor **1108**, and the components of the same, may also be implemented.

[0228] For example, in some embodiments of the direct current motor any suitable type of bearing **1101** may be implemented depending on the design circumstances, intended implementation, environment of application, or the like. Thus, bearings **1101** may be single roller bearings, multiple-roller bearings, thrust bearings, conical bearings, metallic sleeve bearings, or other suitable type of bearing.

[0229] For embodiments where magnetically conductive rotor stack **1104** is mounted in a canted position with respect to shaft **1100**, it may be desirable to include rotational stabilizers **1102** to dynamically balance the rotation of the shaft **1100**. Any suitable stabilizers **1102** may be implemented. For example, in some embodiments stabilizers **1102** may take the form of machined metallic rings containing distributed tungsten weights to achieve dynamic balance, Other configurations are also possible.

[0230] Likewise, in some embodiments, the arbor **1103** may comprise any suitable arbor or mounting mechanism for securing the conductive stack **1104** to the shaft **1100**. For example, in some embodiments, where conductive stack **1104** comprises a laminate stack, it may be desirable to use a compression arbor **1103** that facilitates the securing and positioning of the laminate. Furthermore, arbor **1103** may be formed from any alloy, compound or element which may serve to enhance motor performance, Of course, other arbors **1103** may be implemented depending upon factors such as the type of shaft **1100**, design of the conductive stack **1104**, as well as other factors.

[0231] In some embodiments, magnetically conductive stack **1104** may comprise a stack **1104** of individual disks laminated together. In other embodiments, stack **1104** may comprise a unitary structure, or other similar solid magnetically conductive path. In still other embodiments, stack **1104** may be replaced with any suitable magnetic material that enhances motor performance, including, but not limited to, various steel alloys, various paramagnetic materials, and distributed air-gap materials such as sintered steels and the like.

[0232] Further, in some embodiments the stack **1104** is fashioned to present a substantially cylindrical profile, such as one described with reference to FIG. **12**, thereby ensuring an air gap with the stator of constant, or substantially constant, dimension at the cost of a relatively slight increase in magnetic circuit length. Such an arrangement facilitates a minimum change in magnetic potential energy across the air gap, and the production of a much reduced Speed Voltage (Vs) component of the Back EMF as described herein.

[0233] Likewise, a variety of shaft position sensors **1108** may also be implemented depending upon factors such as motor design, intended implementation, and environmental circumstances. For example, the shaft position sensor **1108** may be comprised of any mechanism capable of generating and sending data to an electronic controller, including, but not limited to multi-quadrant disk encoders with appropriate sensors, slotted disks with optical sensors, magnetic studs with Hall-Effect transducers, metal studs with magnetic proximity sensors, and any other arrangement that may supply necessary information to the controller, either digitally or in analogue fashion. Likewise, some embodiments may locate components of the shaft position sensors **1108** in a variety of locations. For, example, an indicator, sensor, transducer, or other portion of the sensor **1108** may be positioned on the shaft (e.g., shaft **1100**), and may be in communication with other portions of the sensor **1108** located elsewhere. Other position or orientation sensors **1108** are also possible.

[0234] FIG. 23 depicts an axial view of some embodiments of a stator stack **1200** shown in the annular section view of the stack, and including: mounting and alignment holes **1201**, salient pole projections **1202**, coil windings **1203**, and independent coil structures **1204**, either spool-mounted, of free-standing as desired. Dashed line **1205** represents the mean magnetic path for flux manifesting in the annular portion of the stator steel. As also indicated in FIG. **12**, independent coil structures **1204** may comprise a number of windings **1203**. Included in that number of windings **1203** is a surplus amount of windings **1206**. The surplus windings **1206** may be comprised of the additional amount of windings available for a given source voltage and current and due to the reduction of the Speed Voltage component (Vs) of Back EMF caused by the advantageous rotor assembly **1190** design described herein, and which enables the overall flux density produced to remain at the desired amount.

[0235] By way of non-limiting example, a conventionally designed, variable air gap DC motor of a source voltage V and current I may include a number of windings N to produce an output power P for the given V and I. By implementing the Back EMF reducing design disclosed herein, a constant air-gap DC motor can exploit a surplus of windings $N_s>N$ for the same V and I and deliver the same, or greater P. Alternatively, using the concepts disclosed herein, lower values of V and I can be implemented with the Back EMF reducing designs disclosed herein to deliver the same magnitude of P.

[0236] As discussed in connection with FIGS. 20A and 20B, stator poles **1202** and stator stack **1200** may comprise laminations or other material to optimize magnetic flux production without inducing detrimental eddy currents. Other embodiments of the stator assembly, and the components of the same, may also be implemented.

[0237] For some embodiments implementing a multi-pole stator assembly, the stator assembly **1200** may comprise silicone steel laminations, sintered steel alloys, distributed air gap material, or any other material which may suppress the formation of eddy currents and enhance motor efficiency and performance. Further, for some embodiments the stator assembly may have at least four (4), diametrically opposed salient pole projections **1202**, situated at even angular increments around the stator periphery, and aligned in pole pairs 180 mechanical degrees apart, so as to constitute a complete magnetic path through the rotor at all times. Other configurations are also possible. For example, the embodiment shown in FIG. 23 includes six (6) salient pole projections **1202**.

[0238] As discussed, in some embodiments, each salient pole projection **1202** supports an electrical winding or coil **1203** that develops a magnetic field in response to the passage of a DC Current through the winding **1203**. Surplus windings **1206** may likewise be integral with windings **1203** and, likewise, be energized and contribute to the magnetic field. This field provides a magnetic force which acts upon the rotor assembly **1190** and produces a useful torque.

[0239] In some embodiments, the windings **1203** and **1206** supported by said stator salient pole projections **1202**, are inter-connected so as to produce an additive magnetic effect across the entire pole pair, regardless of the magnetic polarity provided by the electronic controller. Other configurations are also possible.

[0240] FIG. 24 is a vertical cutaway view of some embodiments of the motor frame, housing **1300** and stator stack **1200**, and end bells **1301**, but with the entire rotor assembly **1190** left intact for ease of understanding. FIG. 24 illustrates the motor housing **1300**, motor end-belts **1301**, bearing housings **1302**, as well as the relative positions of the motor stator stack **1200**, and the shaft assembly **1100**. Shaft position sensor **1108** is not shown in FIG. 24.

[0241] As shown on FIG. 24, each stator pole (e.g., **1202**A and **1202**D) includes a pole face **1210**. Across the constant air gap from the pole face **1210**, rotor stack **1104** rotates in the region immediately opposite the pole face **1210**. As disclosed herein, the stack **1104** is designed so that, at any given moment in the rotation, the edge of the rotor stack **1104** is opposite a flux zone **1304** located on the face **1210**.

[0242] FIG. 25 shows the apparatus displayed in FIG. 24, except that the rotor assembly **1190** and shaft **1100** have been advanced 90 mechanical degrees, thus demonstrating the maximum angular rotor displacement possible with one pole set energized. As shown, the flux zone **1304** has travelled along the face **1210**. As the rotor assembly **1190** continues to rotate, the flux zone **1304** will travel back-and-forth along the pole face **1210** in a path described by simple harmonic motion.

[0243] FIG. 26 is a functional block diagram of the presently disclosed motor system designed for "Open System Operation," which means that energy recaptured from the motor's inductive components during its operation, will be applied to a capacitive storage element, and utilized to supply power to some electrical load external to the motor itself, such as a lamp, a resistor, a pump, etc. Of course, any suitable external load may be powered in this manner.

[0244] As shown in FIG. 26, the components and general layout of the Open System are as follows. Power incoming to the system from an external source **1500** may be appropriately conditioned and applied to Direct Current Power Supply **1501**. Main Power Storage Capacitors **1502** are also in communication with DC power supply **1501**. Electronic Motor Controller **1503** receives power from DC power supply **1501** and communicates with Motor **1504**. Motor **1504** is driven by controller **1503** and turns a mechanical load **1507**. Of course, mechanical load **1507** may be any suitable load according to the application and implementation. Motor Output Shaft **1505** may correspond to the described embodiments of shaft **1100**. Position Sensor **1506** corresponds to the described embodiments of sensor **1108**. Recapture Capacitor Bank **1508** may receive recaptured power from the motor **1504** via controller **1503** as described in more detail below. Power inverter **1509** can be used to convert the recaptured power to alternating current (AC), for example when powering AC Load **1510**. Unconverted direct current (DC) power from recapture capacitor bank **1508** may be used to power DC Load **1511**. Other configurations of Open System Operation are also possible.

[0245] FIG. 27 is a block diagram of the presently disclosed motor system designed for "Closed System Operation," which means that energy recaptured from the Motor's inductive components during its operation, will be applied to a capacitive storage element and then utilized to send power back to the main power supply by means of a DC to DC converter operating in conjunction with an electronic Feedback Controller.

[0246] As shown in FIG. 27 many components described in connection with FIG. 26 are the same and have similar functionality here. One difference in Closed System Operation is that output from Recapture Capacitor Bank **1508** may be applied to DC to DC Converter **1609** and, through implementation of Feedback Controller **1610**, fed back to primary capacitor bank **1502**. Other configurations of Closed System Operation are also possible.

[0247] FIG. 28 is a block diagram representing some embodiments of the logical control steps occurring within the Electronic Controller which result in the Motor System functioning in the Open System Mode. Again, this means that energy recaptured from the Motor's inductive components (e.g., winding **1203** and surplus winding **1206**) during its operation, will be applied to a capacitive storage element and then utilized to send power to some electrical load external to the motor itself, such as a lamp, a resistor, a puny, etc. Of course, any suitable external load may be powered in this manner.

[0248] As shown in FIG. 28, power incoming into the system from an external source **1700** may be appropriately conditioned and applied to Positive DC Power Supply **1701** and Negative DC Power Supply **1703**. Main Positive DC Capacitor Bank **1702** and Main Negative DC Capacitor Bank **1704** communicate with their respective power supplies. Electronic Controller **1705** communicates with position Sensor **1706**, which corresponds to described embodiments of sensor **1108**. Controller **1705** also functions to power Motor Winding **1707**, which corresponds to the described embodiments of windings **1203** and surplus windings **1206**. Recapture Capacitor Bank **1708** stores the energy from the inductive elements (e.g., windings **1203** and **1206**). External DC Load **1709** may be any suitable load. Power Inverter **1710** may be implemented to condition recaptured energy for application to External AC Load **1711**, which also may comprise any suitable load. In some embodiments Motor Starter **1712** may be implemented to start rotation of the motor as described below.

[0249] FIG. 29 is a block diagram representing the logical control steps occurring within the Electronic Controller which result in the Motor System functioning in the Closed System Mode. As described in connection with FIG. 28, similar components have similar functions. In a Closed System Mode, energy recaptured from the Motor's inductive components (e.g., windings **1203** and surplus windings **1206**) during its operation, will be applied to a capacitive storage element **1708** and then utilized to send power back to the appropriate Positive or Negative Main Power Supply by means of DC to DC converters **1810**, **1811** operating in conjunction with an electronic Feedback Controller **1809**.

[0250] The following is a description of methods of operation for some exemplary embodiments of the presently disclosed system.

[0251] Referring now to FIG. 24, it will immediately be realized, by those skilled in the art, that the application of DC current to pole-coils **1202A** and **1202D** (which include windings **1203** and surplus windings **1206**) will cause the expansion of a DC magnetic field through said pole sets, through the rotor stack **1104** and around the stator mean magnetic path **1205**, such that, the magnetic flux lines will develop a reluctance torque upon the rotor stack **1104**, due to its elliptical shape, and cause a maximum rotor displacement of 90 mechanical degrees, relative to pole pieces **1202A** and **1202D**, to the position illustrated in FIG. 25. However, an angular movement of just a few degrees may be detected by the shaft position sensor **1108** and this information may be sent to the electronic controller **1503**.

[0252] In some embodiments, the controller **1503** may then initiate a timing function, which will allow the rotor stack **1104** to turn through a critical mechanical angle, (e.g., less than 90 mechanical degrees) at which point controller **1503** may cause a DC current to be applied to pole-coils **1202B** and **1202E**, thus locking the rotor at 30 degrees for an instant in time. Simultaneously, the controller **1503** may switch off the current in pole-coils **1202A** and **1202D**, allowing the original magnetic field to collapse down through windings **1203A** and **1203D**, producing a high voltage pulse, and an accompanying current, which the controller **1503** may then direct to recapture bank **1508**.

[0253] The relatively slow collapse of the field lines through pole-coils **1202A** and **1202D**, allows a smooth hand-off of the rotor stack **1104** to the newly energized pole-coils **1202B** and **1202E**, thus completing a total angular displacement of 60 mechanical degrees.

[0254] The charge and discharge rates of the magnetic fields in and through the windings involved shall be a function of factors such as, the particular embodiment's Supply Voltage developed within Power Supply **1501**, the inductance-resistance time constant L/R, the value of the voltage contained within the Recapture Capacitor Bank **1508**, and the impedance of the external load (e.g., **1510** or **1511**.).

[0255] This same switching procedure may be repeated for pole-coils **1202C** and **1202F**, and then again for **1202A** and **1202D**, thereby completing half a rotation, and positioning the rotor stack **1104** properly for the next 180 degree rotation. In some embodiments, the controller **1503** may always supply current of proper polarity so as to prevent reinforcement of magnetic domains within the stator **1200**.

[0256] The next rotation through 180 degrees may be traversed in the same way, re-energizing pole-coils sets **1202B&E**, **1202C&F**, and finally **1202A&D** thereby completing one complete revolution. Each time the controller **1503** switches off a coil set, the resulting collapse of the associated magnetic field will develop an electric pulse which is automatically delivered to the Recapture Capacitor Bank **1508**.

[0257] During normal high speed operation, a continuous stream of electrical pulses will be directed into the Recapture Capacitor Bank **1508**, as shaft **1100** power is being delivered to the mechanical load **1507**. The continuous stream of pulses would ordinarily cause the voltage across the Recapture Bank **1508** to rise to destructive levels if the energy contained therein was not utilized in a constructive fashion. Accordingly, this recaptured energy can be drawn off by application of a DC Load **1511** or an inverter and AC Load combination (e.g., **1509** and **1510**), respectively. The utilization of Recaptured Inductive Energy in a load external to the motor **504** shall be referred to as the Open Power Configuration.

[0258] Referring now to FIG. 27, it will be noticed, that the system configuration for Closed Power Operation is similar to that seen in FIG. **26**, except for the fact Energy stored in the Recapture Capacitor Bank **1508** is drawn down by a DC to DC converter **1609**, then directed back to the Primary Capacitor **1502** by use of a Feedback Control Module **1610**.

[0259] This circuit arrangement allows the DC Motor **1504** to become the load for the Recapture Capacitor Bank **1508**, thereby reusing a significant percentage of the Recaptured Energy, and reducing the power required from the Main DC Power Supply **1501**. Theoretically, this Feedback action may be perfected to the point where the external power need support only the system losses. When this is accomplished, the power drawn by the motor will remain constant, while the external power requirements will diminish in proportion to the power contributed by the Recapture Capacitor Bank **1508**.

[0260] The electronic functions described in accordance with the operation of this Direct Current Motor **1504**, are all directed and synchronized by the controller **1503**. The operational logic of this device is demonstrated in FIGS. **28** and **29**. Of course, variations in the functions required may depend upon the desired effect. FIG. **28** illustrates an arrangement advantageous for Open Power System Configuration, while FIG. **29** illustrates an arrangement advantageous for Closed Power System Configuration.

[0261] System Components **1700** through **1711**, designated in FIG. **28**, and System Components **1800** through **1811**, designated in FIG. **29**, define logical operations employed in the functioning of said Electronic Controller, and are explained in more detail in a related application titled "Controller for Back EMI: Reducing Motor," U.S. patent application Ser. No. _____, filed concurrently.

[0262] Referring now to FIG. **28**, it will be noted that Motor Starter **1712** is mounted upon the motor output shaft **1100**. In some embodiments, normal starting procedure for a DC motor **1504** may involve a starting algorithm, Such an algorithm may be supplied by the controller **1503**, which will pulse the Stator windings (e.g., **1202**) in proper sequence to induce angular speed. However, should the need arise for a separate high-torque starting means, then it may be supplied in the manner illustrated. For example, a shaft-mounted device (e.g., **1712**) utilizing separate starting windings, a starter motor, or any other starting method known to and practiced by the electric motor industry.

[0263] While the invention has been described in detail and with reference to specific embodiments thereof, it will be apparent to one skilled in the art that various changes and modifications can be made therein without departing from the spirit and scope thereof. Accordingly, it is intended that the present invention cover the modifications and variations of this invention provided they come within the scope of the appended claims and their equivalents.

What is claimed is:

1. A direct current motor system comprising:

a stator assembly comprising:

an even number of magnetically conductive salient poles, each salient pole comprising a pole face;

a winding for generating magnetic flux within at least one of the salient poles; and

wherein the salient poles are arranged in pairs located on opposite sides of a central axis and positioned to form a stator cavity with a substantially constant circumference;

a rotor assembly comprising:

a shaft mounted to rotate about the central axis;

a magnetically conductive element mounted to the shaft and shaped so that when rotated about the central axis the magnetically conductive element directs a flux zone along the face of a salient pole in a substantially periodic motion, so that the length of the magnetic flux path formed by the magnetically conductive element and the salient pole varies with the substantially periodic motion of the flux zone; and

wherein the magnetically conductive element comprises an outer edge that when rotated about the central axis circumscribes a path within the stator cavity that is substantially uniformly spaced from each salient pole face thereby forming a substantially constant air gap between the outer edge of the magnetically conductive element and each salient pole face; and

a shaft position indicator for indicating an orientation of the shaft and providing input to a control circuit that periodically energizes the winding for generating magnetic flux which causes the magnetically conductive element to move the shaft in a motoring action about the central axis.

2. The direct current motor system of claim 1 further comprising an electronic controller in communication with the shaft position indicator.

3. The direct current motor system of claim 1 further comprising stabilizers that dynamically balance the rotation of the shaft about the central axis.

4. The direct current motor system of claim 1 wherein the magnetically conductive element is substantially elliptical in shape, and is mounted on the shaft at an angle that is canted with respect to the central axis.

5. The direct current motor system of claim 4 wherein the substantially elliptical shape is describable with reference to a circle with a radius r at an angle θ measured from the center of the circle and in the plane of the circle;

wherein a hypotenuse R, may be drawn at an angle of inclination α from the plane of the circle and at a length given by $R = r \cos \alpha$; and

wherein the perimeter of the substantially elliptical shape is described by rotating R about the full 360 degrees of angle θ about the circle while varying the length of R in accordance with $R = r(\cos \alpha)^{-1} \sin \theta$.

6. The direct current motor system of claim 1 wherein the magnetically conductive element further comprises a laminated structure.

7. The direct current motor system of claim 6 wherein the laminated structure further comprises a laminated stack of individual disks.

8. The direct current motor system of claim 1 wherein the magnetically conductive element further comprises a unitary, non-laminated structure.

9. The direct current motor system of claim 1 wherein the magnetically conductive element further comprises a steel alloy.

THE MEANING OF UNITY IN ENERGY CONVERSION SYSTEMS

10. The direct current motor system of claim 1 wherein the magnetically conductive element further comprises a paramagnetic material.

11. The direct current motor system of claim 1 wherein the magnetically conductive element further comprises a distributed air gap material.

12. The direct current motor system of claim 11 wherein the distributed air gap material further comprises sintered steel.

13. The direct current motor system of claim 1 wherein the magnetically conductive salient poles are constructed so as to minimize eddy currents from flux movement in at least two directions.

14. The direct current motor system of claim 13 wherein the salient poles further comprise:
   a shoe portion; and
   a bottom portion.

15. The direct current motor system of claim 14 wherein the shoe portion further comprises a laminated structure with laminations oriented in a first direction, and the bottom portion further comprises a laminated structure with laminations oriented in a second direction.

16. The direct current motor system of claim 15 wherein the first direction and the second direction are substantially orthogonal.

17. The direct current motor system of claim 14 wherein the shoe portion further comprises a grain-oriented steel structure with a grain oriented in a first direction, and the bottom portion further comprises a grain oriented steel structure with a grain oriented in a second direction.

18. The direct current motor system of claim 17 wherein the first direction and the second direction are substantially orthogonal.

19. The direct current motor system of claim 13 wherein the salient poles further comprise sintered steel material.

20. The direct current motor system of claim 13 wherein the salient poles further comprise ferrite material.

21. The direct current motor system of claim 13 wherein the salient poles further comprise distributed air gap material.

22. The direct current motor system of claim 1 wherein the salient poles are of a size that keeps the overall magnetic circuit length at an optimum value to lessen motor iron losses.

23. The direct current motor system of claim 1 wherein the winding further comprises a number of turns of electrical conductor.

24. The direct current motor system of claim 23 wherein the conductor size and number of turns are at an predetermined amount to establish a magnetic flux of a predetermined value and keep copper losses to a minimum.

25. A direct current motor system comprising:
   a Back-EMF reducing DC motor comprising an energizing coil;
   a sensor that senses an operational condition of the DC motor;
   a recapture storage device that supplies power to an electrical load; and
   a controller that receives input from the sensor relevant to an operational condition of DC motor, controls the energizing of the energizing coil in response to the sensor input, and directs recaptured energy from the energizing coil to the recapture storage device.

26. The direct current motor of claim 25 wherein the DC motor further comprises:

a shaft, and the sensor is a position sensor that provides information to the controller related to the position of the shaft.

27. The direct current of claim 25 wherein the electrical load is an electrical load external to the DC motor.

28. The direct current motor or of claim 25 wherein the electrical load is an electrical load that participates in the supplying power to the DC motor.

29. The direct current motor of claim 28 wherein the controller reduces the energy drawn from an external power source and used to operate the DC motor by an amount related to the energy stored in the recapture storage device.

30. A method for operating a DC motor comprising:
   energizing a first winding located on a salient pole of a stator assembly, wherein the energized winding generates a magnetic flux upon energizing;
   rotating a rotor assembly in response to the magnetic flux, and wherein the rotor assembly includes a magnetically conductive element and wherein the rotor assembly comprises a shaft;
   communicating an orientation of the shaft to a controller;
   energizing a second winding and de-energizing the first winding in response to the communicated shaft orientation; and
   capturing an electrical pulse, generated in the first winding in response to the collapsing magnetic flux associated with the de-energizing of the first winding, in a storage device.

31. The method of claim 30 further comprising:
   communicating a second shaft orientation of the shaft to the controller;
   energizing the first winding and de-energizing the second winding in response to the communicated second shaft orientation; and
   capturing an electrical pulse, generated in the second winding in response to the collapsing magnetic flux associated with the de-energizing of the second winding, in a storage device.

32. The method of claim 31 further comprising:
   accumulating the electrical pulses generated in response to the collapsing magnetic flux associated with the de-energizing of the first and second windings in the storage device.

33. The method of claim 32 further comprising:
   utilizing the energy stored in the storage device as a result of the accumulation of the electrical pulses by applying the energy to an electrical load.

34. The method of claim 33 wherein the electrical load is a load external to the DC motor.

35. The method of claim 33 wherein the electrical load is a load that participates in supplying power to the DC motor.

36. The method of claim 35 further comprising:
   reducing the energy drawn from an external power source and used to operate the DC motor by an amount proportional to the energy stored in the storage device.

37. A stator assembly comprising:
   an even number of magnetically conductive salient poles, each salient pole comprising a pole face;
   a winding for generating magnetic flux within at least one of the salient poles; and
   wherein the salient poles are arranged in pairs located on opposite sides of a central axis and positioned to form a stator cavity with a substantially constant circumference; and

wherein the magnetically conductive salient poles are constructed so as to minimize eddy currents from flux movement in at least two directions.

38. The stator assembly of claim **37** wherein the salient poles further comprise:
 a shoe portion; and
 a bottom portion.

39. The stator assembly of claim **37** wherein the shoe portion further comprises a laminated structure with laminations oriented in a first direction, and the bottom portion further comprises a laminated structure with laminations oriented in a second direction.

40. The stator assembly of claim **39** wherein the first direction and the second direction are substantially orthogonal.

41. The stator assembly of claim **38** wherein the shoe portion further comprises a grain-oriented steel structure with a grain oriented in a first direction, and the bottom portion further comprises a grain oriented steel structure with a grain oriented in a second direction.

42. The stator assembly of claim **41** wherein the first direction and the second direction are substantially orthogonal.

* * * * *

The End

THE MEANING OF UNITY IN ENERGY CONVERSION SYSTEMS

ESTC

Energy Science & Technology Conference

Meet Jim Murray & Other Pioneers of the Modern-Day Tesla Movement

http://energyscienceconference.com

Tesla's Hidden Discoveries

Video

http://teslashiddendiscoveries.com

Dynaflux Alternator

Video

http://dynafluxalternator.com

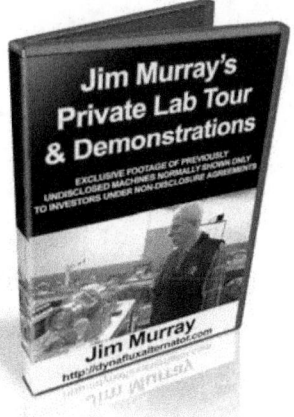

Jim Murray's Private Lab Tour & Demonstrations

Video

http://dynafluxalternator.com

Jim Murray Patent Collection

PDF Document

http://dynafluxalternator.com

The Dynaflux Concept and Lenz's Law

Video

http://dynafluxalternator.com

Fundamentals of the Transforming Generator

Video

http://emediapress.com/jimmurray/tgen

The Secret of Tesla's Power Magnification – Co-Presented by Paul Babcock

Video

http://teslaspowermagnification.com

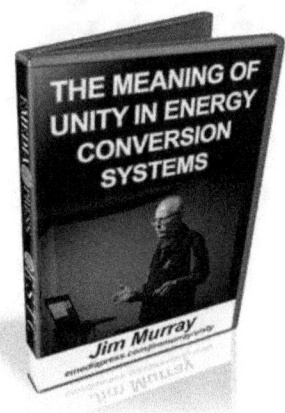

The Meaning of Unity in Energy Conversion Systems

Video

http://emediapress.com/jimmurray/unity

www.teslascientific.com

THE MEANING OF UNITY IN ENERGY CONVERSION SYSTEMS

THE MEANING OF UNITY IN ENERGY CONVERSION SYSTEMS

The Lakhovsky MWO remained a mystery until about 10 years ago because nobody actually knew what circuit he was using. Some originals were found crated up and they were reverse engineered and turned out to be quite different from Lakhovsky's patent drawings.

This is one of the most desired high frequency and high voltage machines ever created and only a few companies are manufacturing them according to Lakhovsky's actual design.

In Europe, you can find them from $15,000 to $25,000 USD – they are very expensive because they are built like replicas of the early 1900's units for those that want the old look.

The only company in the United States making them puts them in a wooden box so it is not shielded and creates a massive amount of interference.

All the "MWOs" that uses plasma tubes, ignition coils and other output methods are not true to the original Lakhovsky design.

Our pulse modulator can be used to power Tesla Coils for regular experimentation, you can use this unit if you want to replicate Eric Dollard's Cosmic Induction Generator invention, etc. This unit is truly **UNIVERSAL** and we are the only ones that are offering this!

If the US Navy, Bell Labs and Western Electric combined their talents to build a Lakhovsky MWO, then this is what you would wind up with – a compact, highly shielded, low loss and high power output unit that is practically military spec. There is no comparison anywhere in the world, not even close.

The antennas are precisely measured to spec and are works of art!

We are constantly updating the design so that the longevity and quality continue to improve - most of these modification are invisible to the user, but it is important to us that any possible enhancement that adds to the experience of the user will be passed on at no extra charge.

THE MEANING OF UNITY IN ENERGY CONVERSION SYSTEMS

A BIT ABOUT THE UNIT WE ARE MAKING AVAILABLE IN LIMITED QUANTITIES AT THIS TIME

Paul Babcock is also on the team of consultants who has assisted in the testing procedure and we have verified that the waveform that our MWO units produce is 100% identical to Georges Lakhovsky! He was the first one of our associates to build a MWO from scratch based on the real circuit used by Georges Lakhovsky.

Paul Babcock got us on the right track.
Peter Lindemann contributed to the design.
Eric showed up and the rest is history.

We will continue to make refinements as needed and we are honored to be able to offer this " RARE " machine that was made possible by these engineers who have inspired and helped us with their genius.

Instead of our units looking like something from the early 1900's, it looks like a modern analog military radio design in a small compact package and it is 100% electrically identical to the Lakhovsky MWO. There is no need for it to come in a case 5 times this size. The Pulse Modulator in the picture on the next page is what we are manufacturing.

The units we are offering you will also NOT have a twist lock plug connection front center as it is not needed, it will have a common trapezoid shaped computer or monitor power supply plug. The left switch is a magnetic-hydraulic circuit breaker that turns on the main power, top center is a dial for the timer and the dial on the right is the VARIAC (Variable AC) dial.

The cases are powder coated (the best kind of painting/finishing method) and silk-screened with white labeling on the front for all dials, etc. along with the name and model of the unit.

You can see on the right side is the spark gap vent and dial to adjust the spark gap.

THE MEANING OF UNITY IN ENERGY CONVERSION SYSTEMS

Vril Science MWO

https://vril.io

A PERCENTAGE OF THE PROCEEDS FROM EVERY SALE ARE DONATED TO EPD LABORATORIES, INC. TO HELP SUPPORT THE WORK OF ERIC DOLLARD

BOOKS & VIDEOS

Premier Source of Pioneers of the Modern-Day Tesla, Breakthrough & Free Energy Movement

CONFERENCE VIDEOS along with over 57 books & video presentations published by the world's leading authorities on the subjects who are the actual pioneers of the modern-day Tesla movement. Double opt in subscribers to newsletters, etc. are over 107,000.

Subscribe free here: http://emediapress.com/energytimes.php
Products here: http://emediapress.com

Affiliates Earn 60%

These digital products are offered through Clickbank and commission for each book or video sold earn you 60%. Join our affiliate program and spread the new insights.

Click here and join the team.
http://emediapress.com/affiliate-program/affiliate-program-signup/

THE MEANING OF UNITY IN ENERGY CONVERSION SYSTEMS

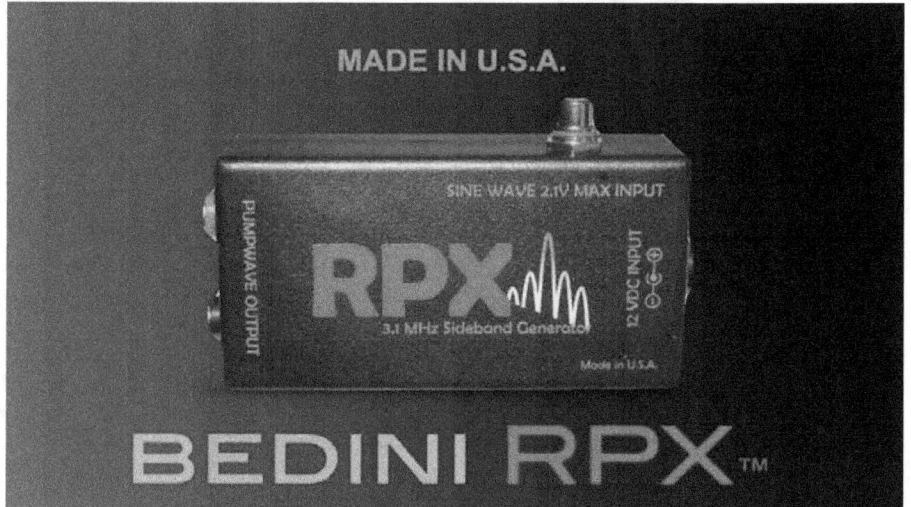

The **BEDINI RPX SIDEBAND GENERATOR** ™ is one of the few devices available to the public that creates the "Rife Frequencies" in the same way that Rife originally conceived.

Many devices available that claim to be "Rife Machines" are simply signal generators and that is it. With the Bedini RPX, a signal generator is used, but it is only one part of several components required to get it right. When audio signals from a function generator are mixed with the high frequency carrier produced in the Bedini RPX unit, they combine to create what are known as sidebands. These frequencies automatically sweep through a wide range and hit every "Rife Frequency" even if those exact frequencies are unknown to the user. And, it happens without having to program in any particular frequency!

Visit our website and watch the free videos on the page so you can see what these sideband frequencies look like and why the do indeed match what Rife himself was creating.

The Bedini RPX is also the least expensive unit available that truly replicates the brilliant work of Royal Raymond Rife. Full systems (combos) and wholesale discounts are now available.

Learn more: https://vril.io

THE MEANING OF UNITY IN ENERGY CONVERSION SYSTEMS

**SOLAR SECRETS
FREE DOWNLOAD RIGHT NOW!**

http://freesolarsecrets.com

THE MEANING OF UNITY IN ENERGY CONVERSION SYSTEMS

**WATCH FREE VIDEOS
PRESENTED BY
TESLA MEDIA NETWORK
ON CONNECTED TELEVISION**

**MANY HOURS OF FOUNDATIONAL PRESENTATIONS BY
THE
TELSA MASTERS**

http://teslamedianetwork.com

ENHANCES ENDURANCE AND REVERSES AGING

ASEA REDOX products are the first and only products on the market that contains active redox signaling molecules, cellular messengers vital in protecting, rejuvenating, and restoring cells. These molecules, native to the human body, are created through a groundbreaking, patented process that reorganizes molecules of natural salt and purified water into redox signaling molecules.

ASEA REDOX Cell Signaling Supplement addresses cellular breakdown, starting at the genetic level. This supplement is created using a groundbreaking, patented process that reorganizes molecules of natural salt and purified water into redox signaling molecules and has been scientifically tested and shown to signal the activation of genetic pathways.

http://index.teamasea.com

THE MEANING OF UNITY IN ENERGY CONVERSION SYSTEMS

www.ingramcontent.com/pod-product-compliance
Lightning Source LLC
Chambersburg PA
CBHW070648220526
45466CB00001B/346